NUMERICAL MODELING OF AAR

Numerical Modeling of AAR

Victor E. Saouma

Department of Civil Engineering, University of Colorado, Boulder, CO, USA

CRC Press
Taylor & Francis Group
Boca Raton London New York

CRC Press is an imprint of the
Taylor & Francis Group, an **informa** business

CRC Press
Taylor & Francis Group
6000 Broken Sound Parkway NW, Suite 300
Boca Raton, FL 33487-2742

First issued in paperback 2018

CRC Press/Balkema is an imprint of the Taylor & Francis Group, an informa business

© 2014 by Taylor & Francis Group, LLC

No claim to original U.S. Government works

ISBN-13: 978-0-415-63697-1 (hbk)
ISBN-13: 978-1-138-07470-5 (pbk)

Library of Congress Cataloging-in-Publication Data

Applied for

Published by: CRC Press/Balkema
                P.O. Box 11320, 2301 EH, Leiden, The Netherlands
                e-mail: Pub.NL@taylorandfrancis.com
                www.crcpress.com – www.taylorandfrancis.com

**Visit the Taylor & Francis Web site at
http://www.taylorandfrancis.com**

**and the CRC Press Web site at
http://www.crcpress.com**

Contents

List of Figures

List of Tables

Preface

Many structures throughout the world are known to be affected by the alkali aggregate reaction (AAR). Once AAR has been identified in a structure, a number of questions must then be answered, namely: how can the time-dependent impact of AAR on serviceability and strength be determined? when would the reaction be expected to stop?; and what would be the eventual impact of remediation steps (such as slicing a dam)?

This book focuses exclusively on the numerical simulation of AAR using exclusively methods developed and applied by the author. These include a macro-constitutive model which could be implemented in an existing finite element code, a micro model to simulate the chemical reaction and a numerical model in order to assess the residual life of a structure affected by AAR. Though it deals mainly with numerical methods, the book also addresses peripheral issues, such as the role of AAR in nuclear reactors and dams, approximate methods to identify residual expansion, reaction kinetics as well as a range of sensitivity issues. On the other hand, this book is not intended to: determine whether or not AAR has occurred, provide solutions to prevent AAR from occurring or remedy its eventual presence, nor would it present any numerical model developed by others.

Even though some of the models provided in this book have already been published elsewhere, much original work is indeed displayed. First through complementary (previously unpublished) validation and parametric studies, then through innovative material such as a mathematical model for AAR's kinetics and an effort to mathematically interpret laboratory test results for the purpose of estimating residual expansion. The main advantage of the constitutive model presented, as opposed to the majority of contemporary models in the literature, is its ease of implementation in conjunction with other authors' linear or nonlinear material constitutive models. Two such implementation performed by other researchers are presented (along with applications undertaken by the author). Herein lies the major advantage of the proposed constitutive model: the ease of its implementation in existing finite element codes.

This book relies extensively on the model developed by Larive and Multon, who more than any other researchers laid the groundwork for a modern understanding of AAR. Accordingly, the author is extremely grateful to them for their breakthroughs in the field. While an excessive reliance on Larive's Equation may be the subject of debate, it has nonetheless successfully provided a mathematical framework for building a physics and mathematics-based model with all of the associated uncertainties. This approach offers sufficient rigor for determining a rigorous estimate (albeit far from perfect) for long-term behavior.

It goes without saying that informed observers will most certainly find pitfalls or shortcomings with some of the models set forth herein, regarding perhaps the set of assumptions, the underlying methodologies or even the conclusions. Yet, to the best of the author's knowledge, a book of this scope and purpose has never before been published, and this is a first timid attempt to fill the gap.

The book has been divided into 9 chapters and 3 appendices. The introductory chapter is very detailed, beginning with a brief review of the meaning of AAR, with its effect on dams and nuclear power plants. An overview of the finite element method (FEM) then offers sufficient detail for the reader to appreciate the flexibility of the author's constitutive model presented in Chapter 2. This introduction also covers the FEM application to heat transfer and concludes by exposing a set of issues surrounding the implementation of constitutive models for AAR using FEM. The second chapter describes in great detail the AAR constitutive model developed by the author. Chapter 3 starts off by addressing the issue of concrete cracking (a natural result of AAR) before presenting two constitutive models for concrete, the first as a nonlinear continuum based on plasticity and the second using nonlinear discrete joint elements based on fracture mechanics. Chapter 4 is dedicated to validating the author's model, while the next chapter serves as an extensive parametric study of this model, equipping the analyst with guidelines on the importance of various parameters and a sense of what are the most important effects to be taken into account in an AAR finite element study conducted today. Chapter 6 focuses on the material properties to be introduced in an AAR investigation, it relies extensively on a field study by the US Bureau of Reclamation. Chapter 7 reviews a number of applications based on the author's model and two others developed by other researchers who adopted it. In deviating from the earlier part of the book, Chapter 8 offers a micro-model based on diffusion models to simulate AAR in a mortar bar. An innovative mathematical model for AAR kinetics has also been inserted here. The final chapter tackles the important issue of residual expansion by seeking to determine when the reaction will actually stop. An innovative model, built in part from mathematics principles, has been included as well. Yet, this is still work in progress.

The first appendix provides a series of numerical benchmark problems designed as a means for finite element code validation. The solutions to these problems were initially given in Chapter 4. Appendix B consists of a very brief description of the finite element code MERLIN, developed by the author and used in the AAR analyses conducted in Chapter 7. The third and last appendix is a quick overview of chemical reaction kinetics and intended to facilitate understanding of Chapter 8.

The author would like to express his deepest thanks to Dr. Luigi Perotti, Dr. Wiwat Puatatsananon, Mr. Mohammad Amin Hariri Ardebili and Mrs. Ruth Martin for their valuable assistance throughout this effort. Sincere acknowledgments are also extended to the Tokyo Electric Power Company (TEPCO) for financially supporting much of the numerical implementation of the author's model into the MERLIN finite element code. The author is very grateful to Dr. Georges Darbre for providing the initial financial support of this work on AAR. A big round of thanks is addressed to Mr. Robert Sachs for a much appreciated copy-editing job on the original manuscript. The review of Section 1.2.10 by Dr. Naus is also hereby acknowledged, and lastly, special commendation goes to the author's colleague Prof. YunPing Xi for his invaluable input over the course of multiple discussions.

Victor Saouma
Boulder, CO 2013

About the author

Victor E. Saouma is a professor of civil engineering at the University of Colorado Boulder. He joined the department in 1984 where he teaches courses in structural analysis. He is currently president of the International Association of Fracture Mechanics for Concrete and Concrete Structures (IA-FraMCoS) and was formerly the director of the University of Colorado Fast Hybrid Testing Laboratory which is part of the George E. Brown, Jr. Network for Earthquake Engineering Simulation.

Over the years his research interests have varied but are always driven by a desire to apply first principles toward the solution of engineering problems. This has included innovative experimental work such as centrifuge/shake table tests of dams and real time hybrid simulation of reinforced concrete frames, as well as development of constitutive models, development of nonlinear finite element codes, modeling of concrete.

His research has primarily been funded by EPRI (Electric Power Research Company), TEPCO (Tokyo Electric Power Company), and government agencies such as the National Science Foundation and the Oak Ridge National Laboratory. As a consultant, his work has involved the seismic safety of very high arch dams, delamination in nuclear power plants, and AAR induced damage in infrastructures. He has over eighty peer-reviewed journal articles.

1 Introduction

This book addresses the topic of numerical modeling of the alkali aggregate reaction (AAR)[1]. As a basic premise, it is assumed that the reader must face a swelling structure problem and answer two questions of paramount importance: 1) what are the structural (and operational) consequences of increased swelling of the structure; and 2) when would swelling stop. In other words, should structural cracks, misalignment of certain critical components and localized failures be expected? Moreover, when should swelling reach a critical stage, what would be the effectiveness of the only proven remedy, i.e. slotting to relieve stresses(applicable only in dams)? This book will not address either of the issues previously raised in hundreds of papers and books: a) how to identify the aggregates and cement that avoid swelling; and b) is AAR occurring in a structure? The answers to those questions can be found in surprisingly few books Institution of Structural Engineers (1992), West (1996), Blight and Alexander (2008), along with countless conference proceedings (such as those of the ICAAR *International Conference on Alkali-Aggregate Reaction* (held since 1974) and, more recently, RILEM and ICOLD reports). Once again, it is important to emphasize that this book's intended readers accept as a given that AAR exists and moreover that its impacts need to be assessed, i.e. petrographic tests are no longer required to determine whether a chemically-induced expansion is present.

This first chapter will provide the reader with a certain amount of background material to better understand the underlying principles of concrete, AAR, heat transfer and finite element method (FEM).

First section will review concrete, then the second will broadly describe AAR: its definition, causes of swelling, onset, and consequences. A succinct description of the most prevalent tests will also be discussed. This will be followed by a brief review of the finite element method (part of which is important to properly grasp the subtleties of the AAR model presented in the next chapter). Because temperature is a key factor governing the AAR expansion, a separate section is devoted to a brief review of heat transfer. Finally,we will address the various approaches to a numerical simulation of AAR through the FEM. Two radically different philosophies have been proposed for this modeling set-up, both of which will be outlined and presented with reference to the major existing models: what are the advantages and disadvantages of each approach, and how is the method presented in this book positioned with respect to other models?

---

[1] Throughout this book, the term AAR will be used of the more restrictive Alkali Silica Reaction (ASR).

1.1 CONCRETE COMPOSITION

AAR occurs when certain siliceous materials[2], occurring in natural mineral aggregates, react with the alkaline components of cement, with the increased molar volume of the products, relative to those of the reactants, creating swelling pressures. Once these pressures can no longer be readily accommodated (by pores for instance) or restrained within hardened concrete, cracking induces expansion.

This expansion (swelling) is a consequence of chemical reactions, while the potential for reaction is itself determined by the chemicals and the mineralogical nature of concrete system components (cement, aggregate, water, etc.) as well as by service conditions (temperature, relative humidity).

Henceforth, for readers unfamiliar with the complexity of concrete, it would be opportune to start by a brief review of concrete, with a special emphasis on those features affecting AAR.

Water is the activating agent for hydration (i.e. the process through which mixed cement and water solidify); it is essential since the precursor binder materials are anhydrous (i.e. contain water), or nearly anhydrous, while the set product consists primarily of a gel-like calcium silicate hydrate (CSH), in which water serves as an essential component. This gel-like hydrate displays a variable composition.

Aggregates used for making concretes are graded according to size, in order to effectively fill space and save cement resources. The chosen aggregate also helps control shrinkage, which otherwise would occur in cement-rich mix designs. Moreover, since hydration is highly exothermic (releases heat), the aggregate acts as a heat sink.

Alkali is always present. AAR occurs rather slowly and, over the period of time required for its action, virtually all of the cement alkali, regardless of its source, is likely to become potentially available for release.

SiO$_2$ comprises about 65% by weight of the accessible portions of the earth's crust. Not surprisingly therefore, SiO$_2$ occurs in both "free" form, i.e. as a solid crystalline oxide in which other essential elements are missing, and silicates (compound SiO$_2$ which contains a silicon-bearing anion). In the latter case, the silicon and oxygen are combined with other elements, the most common of which are aluminum, magnesium, calcium, potassium, sodium, iron and hydrogen. Much of the concern focuses on the free SiO$_2$ phase. In nature, SiO$_2$ mainly occurs as mineral quartz.

water/cement (w/c) ratio is an important design parameter. Though the amount of water required for full hydration of the cement, calculated as a w/c ratio, is only about 0.24, it is typically much greater (in the range of 0.35 - 0.55) in order to facilitate mixing. As a result, considerable water remains even after the cement has essentially been fully hydrated. During the initial hydration stages, the aqueous phase is most abundant and more or less continuous given that it wets the solid-phase grains. As hydration progresses however, the space occupied by the aqueous phase becomes increasingly filled with hydration products, mainly gel-like CSH

---

[2]Much of this section is adapted from the classical work of Glasser and Kataoka (1981)

and $Ca(OH)_2$, and the remaining aqueous phase gradually becomes discontinuous. The cement paste itself is intrinsically porous, and considerable space remains to accommodate this fluid, which is called *pore fluid*.

Macropores are embedded in the concrete, especially those with high w/c ratios. They provide for intimate contact between the pore fluid and hydration products, as well as with aggregate particles. Pore fluid thus serves as a bridge between cement hydration products on the one hand and aggregate on the other. The material transport of soluble species, such as alkali ions, readily occurs through the pore fluid. Henceforth, well vibrated concrete features smaller pores and is thus more likely to be adversely affected by AAR expansion than poorly vibrated concrete, whose pores would have to be saturated by the gel before swelling becomes apparent.

The reaction: Aggregate particles, nominally consisting of SiO_2, are thermodynamically unstable in the cement environment. Reaction commences, thus leading to a decrease in the lowering of the free energy of the system. This reaction, or series of reactions, is accompanied by mass transport of OH and alkali ions: the pore fluid is in intimate contact with cement hydration products as well as with aggregate particles and moreover it serves as the main agent of transport. The details of this reaction, at the aggregate interface of the aggregate, are shown in Figure 1.1.

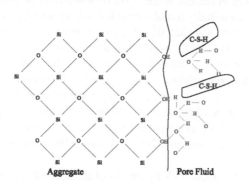

FIGURE 1.1   Aggregate-paste interface, adapted from (Glasser, 1992)

1.2 ALKALI AGGREGATE REACTIONS

Concrete (from the Latin *concretus* which means compact) has been used since the Romans. Hence, many concrete structures have sustained the test of time (such as the Pantheon in Rome), while others have aged at an unanticipated fast rate[3] primarily due to alkali aggregate reaction (AAR).

---

[3]This probably led a great Italian Engineer, colleague of Nervi, to speculate that concrete's reliable life is only 70 years (Galdieri, 2011).

AAR was first identified by Stanton (1940) as a cause for concrete deterioration and is likely the leading cause of dam concrete deterioration. This slowly evolving internal concrete damage causes millions of dollars in damage worldwide, given that no (economically) feasible method is available to stop the reaction. To some extent however, it can be mitigated, as primarily accomplished through an expensive step of slicing the dam to relieve the reaction-induced compressive stresses. Hence, given the need to plan this complex mitigation procedure and keeping in mind that in extreme cases the dam may have to be decommissioned, it has become urgent to provide the engineering profession with sound, practical predictive tools for the evolution in dam structural response. More recently, there has been evidence of AAR in nuclear power plants (see below). This is of major concern, as slicing is obviously not an option.

1.2.1 WHAT IS AAR

Alkali-silica reaction is an acid-base one. The acid reactant is silica in the solid state, the basic one are potassium and/or sodium hydroxide in the pore solution, and calcium hydroxide in the solid state. The reaction medium is water. The product of the reaction is a calcium potassium silicate hydrate, or a calcium sodium silicate hydrate (Dron and Privot, 1992).

Alkalis are present in cement in the form of sodium oxide Na_2, potassium oxide K_2O. These oxides are initially found within the anhydrous phases (i.e. in the absence of water) of the cement, and then dissolve in the pore liquid during the process of hydration into $Na^+ + OH^-$ and $K^+ + OH^-$ forming sodium or potassium hydroxyl ions respectively. Since these ions are not constituents of the cement hydration process, they accumulate in the pore solution. Hence, the alkali themselves do not participate in the reaction, but it is rather their corresponding hydroxyl ions. It should be noted that there is increasing evidence that alkali can also be found in some aggregate where they "cohabit" with silica as long as the pH is below a certain critical value, (Bérubé et al., 2002), (Constantiner and Diamond, 2003). However this presence will be ignored in this book.

Silica (Si) comprises about 65% by weight of the accessible portions of the earth's crust. It occurs both "freeâŁ™ˢ, that is as a solid crystalline oxide from which other essential elements are absent, as well as in silicates, in which silicon and oxygen are combined with other elements, of which aluminium, magnesium, calcium, potassium, sodium, iron and hydrogen are the most important. Hence, silica is the main constituents of most aggregates in the form of silicon dioxide (SiO_2). Whereas the majority of silicon dioxide is stable, the poorly crystallized silica are thermodynamically unstable with respect to the crystalline phase, and prone to react with the cement hydroxyl ions on the surface of the aggregate and produce silanols (Si-OH groups), Fig. 1.1.

Initially, each atom of silicon is connected to the lattice by four siloxane bonds and will be ruptured by OH^- ions. Hence, in the first stage there is a hydrolysis of the reactive silica (siloxane) by OH^- ions to form an alkali-silica gel. In this hydrolysis reaction the high pH pore fluid reacts with Si-O-Si bonds to form silicic acid (silanol bonds) and alkali silicate gel. A characteristic of this reaction is that the dense and impermeable aggregates react very slowly.

$$
\underbrace{- \overset{|}{\underset{|}{Si}} - O - \overset{|}{\underset{|}{Si}} -}_{\text{Siloxane}} + \underbrace{R^+ + OH^-}_{\text{Hydroxyl ions}} \longrightarrow \underbrace{- \overset{|}{\underset{|}{Si}} - O - R}_{\text{Alkali-silicate (gel)}} + \underbrace{H - O - \overset{|}{\underset{|}{Si}} -}_{\text{Silicic acid}} \quad (1.1)
$$

where R^+ denotes an alkali ion such as Na^+, K^+ or Ca^+ (which is dominant at "room temperature", while the former two are prevalent in accelerated tests.

The silicic acid is weak, so that it immediately reacts with further hydroxyl liberating water and negatively charged $Si - O-$, thus readily abundant and mobile sodium, potassium and calcium ions will diffuse in the gel to balance the negatively charged species.

$$
\underbrace{H - O - \overset{|}{\underset{|}{Si}} -}_{\text{Silicic acid}} + \underbrace{R^+ + OH^-}_{\text{Hydroxyl ions}} \longrightarrow \underbrace{- \overset{|}{\underset{|}{Si}} - O - R}_{\text{Alkali silicate (gel)}} + \underbrace{H_2O}_{\text{Water}} \quad (1.2)
$$

The resultant alkali silicate (alkali silicate gel) is hygroscopic (expands in the presence of water), except for $R = Li$,

$$
\underbrace{- \overset{|}{\underset{|}{Si}} - O - R}_{\text{Alkali-silicate (gel)}} + \underbrace{nH_2O}_{\text{Water}} \longrightarrow \underbrace{- \overset{|}{\underset{|}{Si}} - O^- + (H_2O)_n + R^+}_{\text{Expanded alkali silicate (gel)}} \quad (1.3)
$$

Eq. 1.1 to 1.3 follow the notation of Ichikawa and Miura (2007), whereas Chatterji (2005) further addresses the complexity of this chemical reaction.

Different scenarios are possible: a "finite" supply of alkali or silica, or an "infinite" supply of both. In the latter case, the reaction continues for a very long (yet theoretically finite) time, and this has been observed in some dams such as Fontana (US), Chambon (FR) or Salanfe (CH) dams, (Charlwood, 2012). In the former case, the reaction duration is governed by the lower concentration of silica *or* alkali[4]. It is of no concern whether the higher concentration is nearly infinite, as the reaction will eventually cease. Though it is well known that for the reaction to occur, the relative humidity (RH) must be greater than about 80% (Figure 2.11), water must nonetheless diffuse from the pores to enter into contact with the dry gel. Excessive swelling results in cracks, however there is a lack of consensus as to whether cracks initiate inside or around the aggregate; this initiation has no impact on the numerical modeling of AAR at the macro scale.

Finally, an excellent source of information about AAR can be found in the *ASR Reference Center* web page maintained by the Federal Highway Administration (FHWA, 2013) (with emphasis on pavement).

---

[4]This is analogous to an analysis of reinforced concrete beams in flexure. If the reinforcement ratio ρ exceeds the balanced one (ρ_b) then failure is triggered by crushing of the concrete (irrespective of the total area of steel); otherwise, failure is triggered by yielding of the steel (irrespective of concrete strength).

1.2.2   CONSEQUENCES OF AAR

From the standpoint of structural performance, AAR is one of the major causes of deterioration in Portland cement concrete structures around the world. The structures at
high risk of AAR problems are those exposed to wet environments because, as shown
above, one requirement for expansive AAR to occur is moisture. The deterioration,
including expansion, cracking and reduction in engineering properties, of concrete
causes premature distress in concrete structures such as dams, bridges and nuclear
reactors.

Many structures worldwide are affected by alkali aggregate reactions. In many
cases, these impacts can be ignored; for major infrastructure however, such as bridges,
dams or nuclear reactors, it is critical that a reliable safety assessment be performed,
and this step will necessitate a comprehensive numerical simulation.

Maximum AAR expansion is on the order of 0.2-0.5%. This maximum volumetric
strain is equivalent to a thermal expansion with a temperature increase of $200\text{-}500^\circ C$
($\alpha = 10^{-5}$).

When AAR occurs, it has serious consequences at both the structural and material
levels. In the context of this book, the structural consequences are of little significance
since the emphasis herein lies on numerical modeling: we must ultimately address
the effects of AAR on the alteration of mechanical properties.

For reference purposes, at the structural level, AAR will cause cracking (so-called
"map" cracks if no reinforcement is present, and "oriented" cracks in the presence of
internal steel reinforcement). Other effects include: swelling, increase (and possible
rupture) of reinforcement, increase in shear strength, differential swelling (since AAR
is seldom uniformly distributed inside the structure), and popout.

The pattern of cracking (FHWA, 2006) is very similar to cracks generated in masses
of mud when the top layer dries and shrinks or cracks in particular lava flows that occur
when the surface cools rapidly and shrinks. Each portion of the surface pulls away
from every other, generating an irregular honeycomb pattern. The size and regularity
of the pattern depend on the cohesiveness, uniformity, and isotropy of the material and
the speed of shrinkage. This pattern is caused by differential volume change between
the exposed surface material and the attached massive substrate (if any). In drying
shrinkage cracking, or mud cracking, or cooling lavas, the surface has shrunk relative
to the substrate. In expansive AAR, the pattern is generated by an increase in volume
of the substrate relative to the volume of the overlying surface material. The surface
cracking pattern has been called pattern cracking, map cracking[5].

In ordinary concrete that is free to expand equally in all directions parallel with
the plane of the surface, the classical cracking pattern is usually very evident on
the surface in all stages of deterioration, Figure 1.2(a). In continuously reinforced
concrete structures/components that are much longer than they are wide (such as
beams), concrete is not free to expand equally in all directions. In the early stages of
deterioration, concrete can expand only at right angles to the length and the cracks

---

[5]One has to carefully interpret the existence of map cracking as they can be caused by other deleterious
effects such as shrinkage or freeze and thaw.

will, of necessity, be at right angles to the direction of the expansion and therefore parallel with the long dimension (and the reinforcing steel when present), this would lead to directional cracks, Fig 1.2(b).

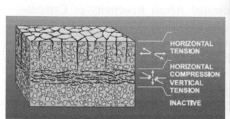

(a) Idealized sketch of cracking pattern in concrete (b) Map cracking with preferred longitudinal trend mass caused by internal expansion as a result of AAR in continuously reinforced concrete pavement

FIGURE 1.2   Cracks generated by AAR (FHWA, 2006)

With regard to the alteration of mechanical properties, most of the literature (Swamy and Al-Asali, 1988a) (Swamy and Al-Asali, 1988b) has focused on newly cast concrete (and too often with a high concentration of reactive aggregates) which is of little interest to readers of this book. Instead, a survey is sorely needed of the deterioration of concrete mechanical properties over many (>20) years. To the best of this author's knowledge, the only such investigation was reported by Dolen (2005, 2011). This report will be addressed separately in Chapter 6.

1.2.3   TESTING METHODS

Laboratory tests can be used either for diagnosis or prognosis. The former seeks to confirm the presence of AAR and determine whether the apparent damage to the structure can reasonably be attributed to AAR. In the context of this book, it is assumed that the structure has been diagnosed, and AAR is present. The later seeks to predict the potential for further deterioration due to AAR. This will receive special attention in Chapter 9.

Broadly speaking, AAR tests fall into one of the following categories:

- Chemical reactivity tests
- Petrographic examination of aggregates.
- Expansion tests to either assess the reactivity of the aggregate only (so-called mortar bar tests where aggregates are crushed and mixed with cement[6]) or reactivity of the actual concrete mix (so-called concrete performance tests where the actual concrete mix is used). In both cases expansion is measured over time.

---

[6]This type of test appears to be derided in (ML12199A295, 2012) *unrealistic specimen preparation.*

- Mechanical tests to assess the deterioration of physical properties with AAR.
- Field observation to measure the concrete crack index (CI).

Addressing testing methods lies beyond the scope of this book, however some will be briefly mentioned. ASTM, (ASTM, 2012) has published the following standards:

C227   Standard Testing Method for Potential Alkali Reactivity of Cement-aggregate Combinations (Mortar-Bar Method).
C289   Standard Test for Potential ASR of Aggregates (Chemical Method).
C295   Standard Guide for Petrographic Examination of Aggregate for Concrete.
C342   Standard Testing Method for Potential Volume Change of Cement-Aggregate Combinations (withdrawn in 2001).
C441   Standard Testing Method for Effectiveness of Mineral Admixtures or GBFS in Preventing Excessive Expansion of Concrete due to ASR.
C856   Standard Guide for Petrographic Examination of Hardened Concrete.
C1260   (AASHTO T303) Standard Test Method for Potential Alkali Reactivity of Aggregate (Mortar-Bar Method).
C1105   Standard Test Method for Length Change of Concrete Due to Alkali-Carbonate Rock Reaction.
C1293   Standard Test Method for Determination of Length Change of Concrete Due to Alkali-Silica Reaction.
C1567   Standard Test Method for Determining the Potential Alkali Reactivity of Combinations of Cementitious Materials and Aggregate (Accelerated Mortar-Bar Method).

Many more have been published in Canada, Japan and Europe (primarily through RILEM). In this context, the excellent work of the French *Laboratoire Central des Ponts et Chaussées* warrants mention, and in particular one of their noteworthy documents on managing structures suffering from AAR in particular, (Divet et al., 2003). Another excellent source is CSA (2000).

1.2.4   CORRELATION BETWEEN TEST RESULTS AND FIELD OBSERVATIONS

In the context of this book, what is more important than the comparability of different accelerated tests is the correlation between laboratory tests and the behavior of on-site concrete (Leemann et al., 2008) for prognosis. Most recently, Leemann and Merz (2012) conducted a study in which both (accelerated) mortar bar tests and so-called concrete performance tests CPT (which test the actual aggregate concrete mix) were performed and then compared with field observation. The accelerated mortar bar test was actually testing the propensity of the aggregate to react by first crushing them, mixing them with cement, and then storing them in an alkaline solution at high temperature. The CPT on the other hand reproduced the actual concrete mix, and also stored at 60°C and 100% relative humidity. Expansion in the former was limited to 20 days, and to the later 20 weeks. In both cases after tests the samples were examined using optical and scanning electron microscopy. Examples of standards for

CPT include ASTM C342-97 (1997), ASTM C227 (2000), ASTM C1293-08b (2008) and AFNOR P18-454 (2004).

Reproducing the concrete of damaged structures, testing it with the CPT and analysing it with microscopy lead to the following observations:

- Expansion rates determined with crack measurements on structures in advanced state of AAR and expansion in the CPT show a good correlation.
- The same rock types react frequently (above average) respectively rarely (below average) in on-site and lab concrete.
- The chemical composition of the reaction products is very similar in both on-site and lab concrete. Non-reactive and potentially reactive concrete mixtures can be distinguished with the CPT. Furthermore, the determined expansion can give an indication about the degree of expansion possible in the structure.

thus "validating" the accelerated test.

Henceforth, to assess the residual expansion of concrete, the concrete performance test (i.e testing during an extended period the actual concrete mix rather than the propensity of aggregates to react) is far better than mortar bar tests.

1.2.5 *IN-SITU* MEASUREMENT: CRACK INDEX

A commonly used (and at time misused) indicator of AAR is the so-called Crack Index (CI) which is a crack mapping process that consists in the measurement and summation of crack widths, Fig. 1.3, along a set of lines drawn perpendicularly (i.e., parallel and perpendicular to the main restraint(s)) on the surface of the concrete element investigated (FHWA, 2010). The method gives a quantitative assessment of the extent of cracking in structural members. In order to generate a statistically representative assessment of the extent of cracking through the CI method, a minimum of two CI reference grids, 0.5 m (20 in) in size, should be drawn on the surface of the most severely cracked structural components. Those components generally correspond to those exposed to moisture and severe environmental conditions, as well as those where ASR should normally have developed to the largest extent.

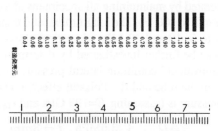

FIGURE 1.3   Simple gage to visually measure crack widths

Yet, one should be very careful in not placing excessive value on the CI as surface cracks may be caused by a multitude of factors other than AAR such as shrinkage, freeze-thaw, carbonation, or presence of reinforcement (which could severely limit cracking in the presence of AAR). Furthermore, surface cracks do not reflect the internal crack/damage as the RH internally is likely to be higher inside the concrete mass than on the surface where cracks are observed and measured. Finally, stress redistribution as a result of AAR is likely to greatly complicate interpretation of surface crack indexes. Nevertheless, it should be mentioned that some correlation between CI and expansion has been reported (Leemann and Merz, 2013).

1.2.6 LCPC EXPERIMENTAL WORK

By far, the most widely cited AAR tests of late have been conducted at the *Laboratoire Central des Ponts et Chaussées* (LCPC) in Paris. These tests, (Larive, 1998b), (Multon, 2004) actually sparked the development of "modern" computational tools for AAR, and models went from "empirical" to "chemo-mechanical" (Section 1.5). Since the model developed in Chapter 2 is very much inspired by LCPC tests, they will be succinctly reviewed.

Larive (1998b)'s monumental work laid the foundation of our modern understanding of AAR. Only those contributions relevant to this book will be reported.

First, the effect of uniaxial confinement was determined by testing many samples with either free or confined expansion. Though it has already been known that a confinement of approximately 8 MPa impeded expansion, this constituted the first systematic investigation. At 20 MPa of confinement, practically no swelling occurs and swelling reaches its peak (0.03%) at a confinement between 5 and 10 MPa (as opposed to 0.2% without confinement).

Another major contribution was that for the first time, the kinetics of expansion were mathematically derived and then the parameter curve could be fitted with experimental data. This curve is shown in Figure 2.1(a).

Up until then, reactive concrete specimens have been subjected to just uniaxial confinement. Multon (2004) investigated the response of reactive concrete cylinders to a multiaxial constraint. 240 by 130-mm cylinders were "wrapped" by stainless-steel rings (3 or 5 mm thick), to provide lateral confinement, and then subjected to constant longitudinal stress by means of creep devices (Figure 1.4). The development of AAR has been accelerated by maintaining all specimens at a constant temperature of 38°C (which corresponds to 100°F). The longitudinal and transversal deformations for cylinders subjected to 0, 10 and 20 MPa of vertical stress are shown in Figure 1.5. These test results can be further investigated by determining (according to thin cylinder theory) the "equivalent" confining lateral pressure due to the cylinder. By neglecting both flexure in the tube and the Poisson effect, it can thus be shown that the equivalent lateral pressure is (assuming E=193 GPa and ϵ_{lat}=0.04%)

$$\sigma_{eq}^{lat} = 2E\frac{\Delta D}{D^2}t = \begin{cases} 3.6\text{MPa} & t = 3mm \\ 5.9\text{MPa} & t = 5.0mm \end{cases} \tag{1.4}$$

Under triaxial confinement, i.e. Figures Figure 1.5(c), 1.5(d), 1.5(e), and 1.5(f) (keeping in mind that longitudinally there is an imposed stress and laterally a confined expansion), a complex expansion occurs; moreover, it can be observed that for

FIGURE 1.4   Triaxial cell used by Multon (2004)

highly-confined specimens, the magnitude of the deformation remains small enough, and we barely have a sigmoid expansion curve (shape predicted by Larive). Longitudinal AAR-induced deformations do not seem to depend on ring thickness (or lateral confinement): for a given ring thickness, AAR-induced axial expansion is higher for a 20 MPa-applied axial stress than for 10 MPa, yet remains significantly less than that obtained with no confinement. These preliminary results are strongly affected by creep (as deduced from measurements), whose magnitude is about 3 times that of the AAR-induced result.

The tabular values of strains extracted from the above figures are shown in Tables 1.1 and 1.2. Let's note that the overall volumetric strains are relatively constant. The effect of stress confinement is perplexing since for a given lateral constraint, the volumetric expansion is first reduced and then increases. This effect may be attributed to microcracking. In contrast, for a given lateral confinement, the longitudinal stress constraint tends to limit the volumetric constraint.

A major conclusion of this work is that an expansion transfer in AAR, as originally revealed by Larive (1998b), is present and the total volumetric AAR-induced strain appears to be constant irrespective of confinement. In other words, the expansion is greatest in the direction of "least resistance". This is further confirmed by field tests reported in Section 6.5.1.1.

Yet, these results are merely preliminary and based on a "Hoek cell"; moreover the confinement is not based on true stress. As such, the author has built an apparatus to subject cubes of concrete to constrained expansion under controlled temperature and humidity. Figure 1.6 shows the testing apparatus[7]. This is currently in progress in collaboration with the Polytechnic University of Catalonia (Prof. Carol).

[7]Development of the AAR triaxial test apparatus was funded by the Tokyo Electric Power Company (TEPCO).

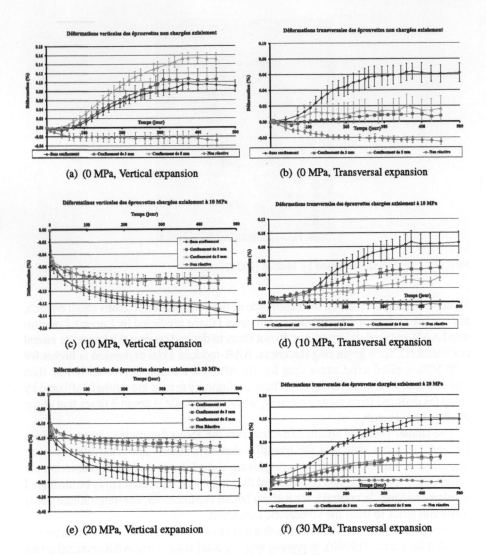

(a) (0 MPa, Vertical expansion

(b) (0 MPa, Transversal expansion

(c) (10 MPa, Vertical expansion

(d) (10 MPa, Transversal expansion

(e) (20 MPa, Vertical expansion

(f) (30 MPa, Transversal expansion

FIGURE 1.5   Selected experimental results of Multon (2004)

1.2.7   PARTIAL FIELD VALIDATION OF LCPC TESTS

Herrador et al. (2008) presented a very interesting study in which tests of large concrete cores drilled from two locations in Belesar dam (Spain) suffering from AAR were performed. Belesar dam is a 600 m long and 129 m high arch dam built between 1957 and 1963. The first series consisted of cores recovered from the top of the dam and the second near the abutments. The former is considered to have been relatively stress free while the second has sustained confining stresses as a result of the expansion. It was also determined that both concrete were nearly identical and had a similar

TABLE 1.1

Longitudinal, Lateral and Volumetric AAR Induced Expansion in terms of Longitudinal Stress Constraint, and Lateral Expansion Constraint, (Multon et al., 2004).

	0 MPa			10 MPa			20 MPa		
Conf. (mm)	0	3	5	0	3	5	0	3	5
P_{eq}^{lat} MPa	0	3.6	5.9	0	4.5	5.9	0	5.3	7.4
Longitudinal	0.12	0.13	0.17	0.0	0.04	0.04	0.0	0.09	0.09
Lateral	0.09	0.04	0.04	0.09	0.05	0.04	0.13	0.06	0.05
Volumetric	0.29	0.21	0.25	0.18	0.14	0.12	0.26	0.21	0.19

Note those are approximate values taken from graphs

TABLE 1.2

Volumetric AAR Induced Expansion when only one constraint is active (Multon et al., 2004).

	Free Expansion	10 MPa	20 MPa	3mm 0 Mpa (3.6 MPa)	5 mm 0 MPa (5.9 MPa)
Volumetric	0.29	0.18	0.26	0.21	0.25

Note those are approximate values taken from graphs

temperature histories. Due to large aggregate sizes, cores ad a nominal diameter of 200 mm and a length of 520 mm.

Specimens were tested *in-situ* inside the dam gallery to maintain a constant temperature and humidity using a a special steel creep frames. Specimens were subjected to constant axial compressive stress and expansion monitored. Three test protocols were pursued: a) free expansion; b) constant stress; and c) variable stress for each of the two test sites.

This research yielded the following conclusions:

1. The expansion free curves in the stress free section of the dam exhibited a nearly linear (residual) expansion of 0.3266×10^6/day. Whereas cores recovered from a confined section of the dam expanded at a rate 3 times higher with a tendency to slow down following a parabolic trend.
2. Testing on creep frames at 2.6, 5.1 and 10.2 MPa constant axial confinement show that in AAR concrete submitted to small compressive load the positive expansion can override the negative creep-induced strains (Fig. 1.7 explains this potential effect)
3. Expansion is mitigated by increased axial confining stresses, beyond 10.2 MPa all expansion is prevented.
4. Under variable stress increase, results can vary, as the expansion may be faster or slower than the rate of mechanical load application.

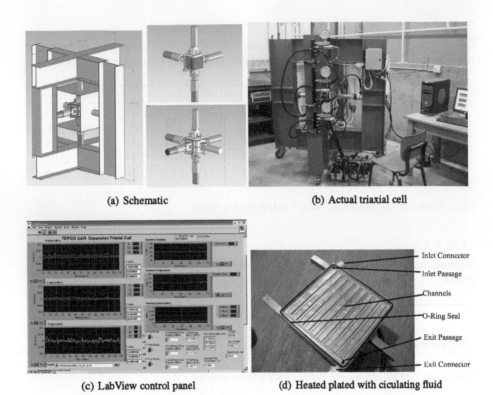

(a) Schematic   (b) Actual triaxial cell

(c) LabView control panel   (d) Heated plated with ciculating fluid

FIGURE 1.6   AAR Triaxial test apparatus

5. In all cases, concrete obtained from the confined area shows more intense expansive behavior than the one extracted from the load-free area.

This experimental study, conducted with actual concrete cores corroborate laboratory derived conclusions about the effect of confinement (Multon, 2004). Furthermore, the increased expansion upon release of the confinement validates the assumption of the author's model in which AAR expansion is to be considered as a volumetric one (Sect. 2.6).

1.2.8   AAR AND CREEP

External manifestation (expansion) of AAR may be hidden by creep and shrinkage (contraction). Assuming a 4,000 psi concrete and a 1,000 psi compressive strain in a dam, the elastic strain is approximately 0.0003. A creep coefficient of 3 will add a creep strain of about 0.001. On the other hand, assuming that AAR initiates immediately, with a $\varepsilon(\infty) = 0.5\%$ or 0.005, then the net strain will be as shown in Fig. 1.7. It should be noted that AAR expansion is plotted with respect to the -0.0003 instantaneous elastic strain. It is now apparent that the external manifestation of AAR has been substantially delayed by shrinkage/creep, and great care should be exercised in discerning shrinkage/creep contraction strains from AAR expansion ones.

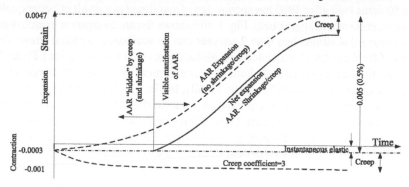

FIGURE 1.7   AAR and Creep Interaction

1.2.9   AAR IN DAMS

Numerous dams worldwide (Charlwood, 2012) have sustained damage from AAR. A list of examples includes: Fontana dam, in the United States (Wagner and Newell, 1995), Mactaquac dam in Canada (Gilks and Curtis, 2003), Canning dam in Australia (Shayan et al., 2000), Chambon dam in France (Peyras et al., 2003), in Iran Jabarooti and Golabtoonchi (2003), Pian Telessio in Italy (Bon et al., 2001), Pracana in Portugal Portugese National Committee on Large Dams (2003), Isola in Switzerland (Malla and Wieland, 1999).

In a recent study in France (which, as we have previously seen, has taken the lead in AAR research worldwide), Sausse and Fabre (2012) reports of a survey conducted by *Electricité de France* (EdF) on 150 dams as shown in table 1.3. One notes that about

TABLE 1.3

Concrete dams operated by EdF, (Sausse and Fabre, 2012)

Type	Monitored	Swelling detection capability	Swelling detected
Gravity	64	24	13
Arch	50	46	15
Gate	20	12	1
Other	22	9	1
Total	156	91	30

20% of the dams have some sort of swelling (none of the dams not monitored did exhibit external signs of distress). Whereas it is commonly assumed that a swelling (double curvature) arch dam will exhibit upstream irreversible displacement, Fig. 1.8(a) confirms this observation, yet single curvature arch dams will expand in the downstream direction. Age is also an important factor, Fig. 1.8(b) shows that a much greater percentage of dams built prior to 1960 exhibit vertical swelling. Indeed, none of the 50 dams built after 1960 has shown sign of swelling (as a result of better quality control on cement and aggregates). Fig. 1.8(c) shows the arc elongation in arch dams. It is assumed that strains less than 2 $\mu\varepsilon$/year can be attributed to global irreversible heating of the dam (it self associated with global decrease in pool elevation). From this figure, we thus infer that 10 arch dams (without vertical displacement devices) have an arc elongation greater than $2\mu\varepsilon$/year. Finally, in Fig. 1.8(d) attention is focused on 11 dams and vertical swelling rate measured during the past 10 years is contrasted with historical measurement record. Only in three cases we observe a decrease in the medium yearly swelling rate during the past 10 years as compared with the historical record. Two of them did indeed benefit from some sort of retrofit.

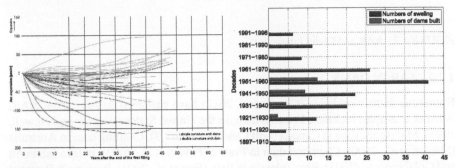

(a) Single vs double curvature swellings in arch (b) Swelling dams in terms of construction date dams   (μm/m/year)

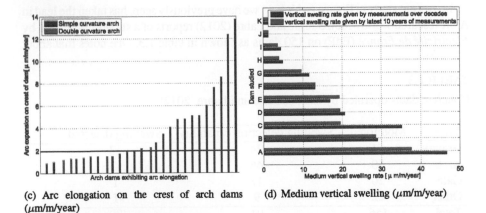

(c) Arc elongation on the crest of arch dams   (d) Medium vertical swelling (μm/m/year)
(μm/m/year)

FIGURE 1.8 Summary of EdF investigation on swelling dams, (Sausse and Fabre, 2012) (b, c, and adapted from)

Finally, it should be noted that from a structural point of view, AAR will cause cracks in the abutments (mostly in V shaped ones) du to the relatively unconstrained

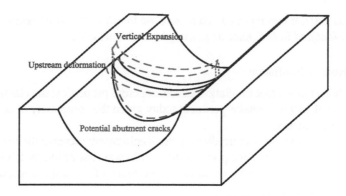

FIGURE 1.9 Abutment cracks due to AAR

vertical expansion, Figure 1.9. Hence, one should be attentive to monitor such potential crack locations.

1.2.10 AAR IN NUCLEAR POWER PLANTS

1.2.10.1 Structural Deterioration

Despite the lack of publicity, some nuclear power plants (NPP) reactors are starting to show signs of Alkali-Aggregate Reactions (AAR).

In Japan, the (reinforced concrete) turbine generator foundation at Ikata No. 1 NPP (owned by Shihoku Electric Power) exhibits AAR expansion and has thus been the subject of many studies. Takatura et al. (2005a) reports on the field investigation work underway: location, extent of cracking, variation in concrete elastic modulus and compressive strength, expansion in sufficient detail to adequately understand the extent of damage. The influence of AAR on mechanical properties (in particular, the influence of rebar) and on structural behavior has been discussed by Murazumi et al. (2005a) and Murazumi et al. (2005b), respectively. In the latter study, beams made from reactive concrete were tested for shear and flexure. These beams were cured at 40°C and 100% relative humidity for about six weeks. Some doubt remains, however, as to how representative such a beam is for those NPP where AAR has been occurring for over 30 years. A study of the material properties introduced in the structural analysis was first reported by Shimizu et al. (2005b). An investigation of the safety margin for the turbine generator foundation has also been conducted (Shimizu et al., 2005a). Moreover, vibration measurements and simulation analyses have been performed (Takatura et al., 2005b). Takagkura et al. (2005) has recently reported on an update of the safety assessment at this NPP.

In Canada, Gentilly 2 NPP is known to have suffered AAR (Orbovic, 2011). An early study by Tcherner and Aziz (2009) actually assessed the effects of AAR on a $CANDU^{TM}$ 6 NPP (such as Gentilly 2). In 2012 however, following an early attempt to extend the life of Gentilly 2 until 2040 (with an approx. $1.9B overhaul), Hydro-Quebec announced its decommissioning after 29 years for economic reasons.

As late as 2007, it was reported that *to date, no incidences of ASR-related damage have been identified in U.S. nuclear power plants* (Naus, 2007).

1.2.10.2 Role of Irradiation

It has long been known that irradiation affects concrete properties; the classical work by Hilsdorf et al. (1978) remains pertinent today given the complexity of conducting supportive experiments.

The possibility that nuclear irradiation can significantly increase the reactivity of silica-rich aggregates (hence the potential for AAR) was first raised by Ichikawa and Koizumi (2002). Let's begin by the author's summary of the state of knowledge in 2002:

- Gamma rays do not affect concrete properties up to a dose of 10^{10} Gy)[8].
- Irradiation of fast neutrons to a dose of more than $\sim 10^{19}$ n/cm^2 results in deterioration of concrete. Irradiation causes the aggregates to expand and the cement paste to shrink.
- The degree of shrinkage at a dose of 5×10^{19} n/cm^2 is about 2% and 0.3% for Portland and alumina cement, respectively.
- The degree of aggregate expansion strongly depends on the type of aggregates. Expansion at a dose of 5×10^{19} n/cm^2 is roughly 1% for limestone and flint, while 0.1% for serpentine.
- The expansion of concrete composed of Portland cement paste and aggregates is greater than that estimated from the volume ratio of aggregates and the degrees of expansion and shrinkage.
- The difference between the estimated and observed degrees of expansion strongly depends on the mineral composition in the aggregates. This difference is higher for aggregates with higher SiO_2 content).
- The difference for concrete with silica-rich flint is about five times higher than that with limestone.
- Concrete with quartzite sand shows severe expansion after irradiation of 3×10^{19} n/cm^2, though crystalline quartz reveals only a small expansion. The degradation in concrete mechanical properties due to fast neutron irradiation is greater for concrete showing a higher degree of expansion.
- The decrease in tensile strength due to irradiation is more pronounced than the decrease in compressive strength.

These results indicate that the degradation of concrete by fast neutrons is not simply due to the individual deterioration of cement paste and aggregates, but rather to complex reactions of irradiated cement and aggregates.

---

[8] One Gray, Gy, is the absorption of one joule of energy, in the form of ionizing radiation, per kilogram of matter, i.e. 1 Gy = 1 J/kg=100 rads

It is thus suggested that the decrease in the resistance to nuclear radiation by increasing the SiO_2 content in aggregates provides a strong indication that the deterioration is due to an acceleration of the alkali-aggregate reaction caused by nuclear radiation.

Based on these premises, the author conducted an experimental program, which led to concluding that:

1. Nuclear radiation significantly increases the reactivity of silica-rich aggregates to alkali. This does not necessarily imply however that AAR will occur since other requisite conditions may not be present.
2. Concrete surrounding the pressure vessel receives the highest radiation dose; in assuming that the lifetime of a commercial nuclear power plant is 60 years, the integrated absorbed dose of the concrete is about 10^9 Gy for gamma rays, which is much lower than the critical dose.

Though not specifically addressing AAR, the work by Vodák, F. and Trtík, K. and Sopko, V. and Kapičková, O. and Demo, P. (2004) is worth reporting. In focusing on the impact of irradiation on the strength of an NPP concrete, the authors investigate the overall effect of irradiation of concrete. They determined that when the concrete used in the construction of Temelin's NPP (Czech Republic) was exposed to γ-irradiation of 6×10^5 Gy (6×10^7 rads) (i.e. which corresponds to 57 years of normal NPP operation time), a degradation in mechanical properties had occurred. More specifically, results indicated that the compressive, splitting-tensile and flexural strengths of the concrete decreased with the dosage level, reaching reductions of about 10%, 5% and 5%, respectively, at the maximum dose. Of more relevance to the topic of this book however is the finding that irradiation generates a succession of chemical reactions, leading to a decrease in the size of pore space and hence inhibiting the ability of concrete to absorb some of the AAR gel produced prior to expansion.

A more recent study by Ichikawa and Kimura (2002) examined the effect of electron-beam irradiation on the reactivity of plagioclase (generally present in volcanic rock) to AAR. It was determined that irradiation of plagioclase with a 30-keV electron beam at a dose in excess of 0.9×10^9 Gy converts a crystalline plagioclase to an amorphous one 35 times more reactive to alkali than the crystal. This observed high reactivity indicates that the deterioration of irradiated concrete by alkali-silica reaction is indeed possible even for aggregates that would otherwise be inert to AAR without irradiation.

In summary, indications suggest that radiation effects on concrete degradation are minimal for the first 40 years; however, a structural life extension to 60-100 years may prove problematic and moreover the data to fully support this concern are insufficient, (Fujiwara et al., 2009).

1.2.10.3 Life Extension

According to the Atomic Energy Act of 1954, (NUREG-0980, 2013) and NRC regulations, the operating licenses for commercial power reactors are issued for 40 years and can be renewed for an additional 20 years, with no limit to the number of renewals.

The original 40-year license term was selected on the basis of economic and antitrust considerations rather than technical limitations. Henceforth, many plants have sought (and obtained) a 20-year life extension. In the United States, the average structural life is 32 years (U.S. Energy Information Administration, 2013).

Also in the U.S., most NPP have already had a life extension from 40 to 60 year, and serious consideration is now being given to a further extension to 80 years. It is in this context that Graves et al. (2013) offered a comprehensive evaluation of potential aging-related degradation modes for light-water reactor materials and components. This work was based on the levels of existing technical and operating experience, knowledge, the expected severity of degradation, and the likelihood of occurrence. The report produced thus detailed an *expanded materials degradation analysis* of the degradation mechanisms capable of affecting concrete.

The report concluded that three (of the five) high-ranked degradation modes could potentially affect the concrete containment, which is the safety-related structure of primary interest.

Creep of the post-tensioned concrete containment.

Irradiation of concrete (which, as shown below, may accelerate AAR).

Alkali-silica reaction *Though this degradation has been well documented through operating experience (for bridges and dams in particular) and the scientific literature, its high ranking in this EMDA analysis underscores the need to assess its potential consequences on the structural integrity of the containment.*

A similar study was conducted in France, a country that contains 58 NPP. This group of plants has an average age of approximately 27 years, spanning a range of 13 to 34 years. Once again, the focus herein lies in the operations of NPP beyond 40 years, and possibly 60 years (Gallitre et al., 2010). With structural aging thus being a concern, it has been recognized that AAR is indeed one of the leading causes of aging-related problems. AAR prevention however has not historically been subject to special requirements in the design and construction of French NPP, and all recommendations for preventing these disorders date back only to 1994 (Coppel et al., 2012). To assess the "theoretical" AAR risk, a methodology was developed by the EdF electric utility in the 1990's based on three parameters:

1. Calculation of active alkali content in the concrete mix according to French LCPC 1994 recommendations. The active alkali content of concrete comprises active alkali from all concrete components (cement, aggregates, water, etc.). This calculation yields the following for the classification of concretes:
   - A1: active alkali content < 2.2 kg/m^3 of concrete
   - A2: active alkali content between 2.2 kg/m^3 and 3 kg/m^3 of concrete
   - A3: active alkali content > 3 kg/m^3 of concrete.
2. Aggregate qualification according to the French LCPC 1994 recommendations. Aggregate qualification is performed from original quarries used for the given site construction. Aggregate qualification yields the following in terms of aggregate classification:
   - NR: Non-reactive aggregate

- PR: Potentially reactive aggregate
- PRP: Potentially reactive with pessimum effect (flint, chart)
3. The environmental description of the concrete structure under consideration includes:
   - H: Humidity exposure (relative humidity > 80%)
   - T: Temperature exposure (T>35°C)
   - N: Normal exposure (T<35°C and relative humidity < 80%)
   - A: Alkali exposure (from a system containing alkalis)
   - M: Marine or industrial water exposure

The risk of AAR is then determined from Table 1.4.

TABLE 1.4

"Theoretical risk of AAR in NPP, (Coppel et al., 2012)

Active Alkali Content	A1<2.2kg/m³					A2 >2.2kg/m³ A2 <3kg/m³					A3>3kg/m³				
Environ.	N	H	T	A	M	N	H	T	A	M	N	H	T	A	M
NR	0	0	0	0	0	0	0	0	0	0	0	1	1	1	1
PR	0	1	1	4	2	1	3	3	4	3	1	4	4	4	4
PRP	0	0	0	2	1	0	x	x	2	1	x	1	1	2	1

"Theoretical" risk of AAR: 0, Negligible; 1, Low; 2, Medium; 3, High; 4, Very high; x, Specific pop-out risk

All 58 NPP in France have been assessed for potential AAR expansion, and only four out of a total of 19 are considered to be at potential risk; these four are essentially located on the Loire River, where aggregates are known to contain a relatively high percentage of silica. For cases of serious concern, practically no solution exists; however, since this phenomenon progresses slowly, it can be detected through observation (such as pop-out or concrete expansion) when the structure has been instrumented, (Gallitre and Dauffer, 2010). If ASR is detected, then its evolution can reasonably be monitored through crack length measurements inside a fixed area using the so-called "trihedron" method. One of three conclusions is deemed possible (Gallitre and Dauffer, 2010):

- In the medium term, no dangerous consequences arise as a result of the large amount of reinforcement in the concrete.
- In a longer term, inspections may be intensified should signs be detected, and more sophisticated analysis may be undertaken if necessary.
- In the long term, tightness may be improved, if necessary, by means of either a metallic or organic coating.

The author would disagree with these conclusions since a solid conclusive understanding of the evolution of this reaction and its impact on NPP structural integrity is necessary *before* granting a license extension.

1.2.10.4 Seabrook Nuclear Power Plant

This section will provide detailed information on the first reported nuclear power plant in the U.S. known to suffer from AAR. All information reported has been gathered exclusively from the Agencywide Document Access and Management System (ADAMS), the official record-keeping system through which the U.S. Nuclear Regulatory Commission (NRC) provides access to publicly available documents.

Description NextEra Energy Seabrook, LLC submitted an application for renewal of the Seabrook Station NPP Unit 1 operating license for another 20 years (beyond the current licensing date of May 15, 2030), (ML12160A374, 2012). This renewal process consisted of two concurrent reviews, i.e. a technical review of safety issues and an environmental review. For the safety review, the License Renewal Rule process and application requirements for commercial power reactors are based on two key principles: a) that the current regulatory process, continued into the extended period of operation, is adequate to ensure that the continuing license basis of all currently operating plants provides an acceptable level of safety, with the possible exception of the detrimental effects of aging on certain systems, structures, and components (SSCs), and possibly a few other issues related to safety only during the period of extended operation; and b) each plant's continuing license basis is required to be maintained.

As part of the license renewal process, an aging management program (AMP) is to be identified that is determined to be acceptable to manage potential problems such as AAR.

In 2009-2010, it was determined that groundwater infiltrated into the annular space between the concrete enclosure building and concrete containment. The bottom 6 ft of the concrete containment wall was in contact with groundwater for an extended period of time. Cracks due to the alkali-silica reaction (ASR) had been observed in various Seabrook plant concrete structures, including the concrete enclosure building ((ML12160A374, 2012), Fig. 1.10). As a consequence, the NRC identified ASR as an open item indicating that it had not been adequately addressed in the Structures Monitoring AMP (OI 3.0.3.2.18-1).

A total of 131 cores (4" diameter, 14" - 16" deep) in the affected areas were tested to determine their compressive strength and modulus of elasticity and then compared with test results from standard concrete cylinders cast during the original concrete construction placements. In addition, petrographic examinations, as per ASTM C856, were performed. It was determined that the areas affected were highly localized, and core samples extracted from adjacent locations did not show signs of ASR. Furthermore, when the core lengths were evaluated (i.e. depth into the wall), it was observed that cracking was most severe at the exposed surface and reduced towards the center of the wall, (ML12199A295, 2012). As a consequence, the NRC initiated an *Open Item* (OI 3.0.3.2.18-1) related to the AMP.

 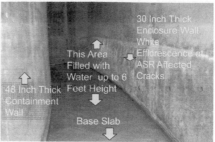

(a) Observed Map Cracks   (b) Identification of damage

FIGURE 1.10   AAR in Seabrook nuclear power plant, (ML12199A300, 2012)

As a result of this identification of AAR, it was reported that NRC officials informed the power plant's owners that in order for the plant to gain approval for its license extension, proof needed to be provided concerning the impact ASR will have on the plant as it ages, as well as the steps adopted to mitigate ASR in the plant's concrete structures, if necessary, (Chiaramida, 2013). Moreover, the NRC made it clear that a final decision on the license renewal application would not be announced until concrete degradation issues identified at the plant had been satisfactorily addressed, (Haberman, 2013).

Root cause investigation   A root cause investigation was performed and led to determining that (ML13151A328, 2013):

RC1 - ASR developed because the concrete mix designs unknowingly utilized a coarse aggregate that, in the long term, would contribute to the Alkali Silica Reaction. Although testing was conducted in accordance with ASTM standards, these standards were subsequently found to be limited in their ability to predict slow reactive aggregates that produce ASR in the long term.

RC2 - Based on the long-standing belief that ASR is not a credible failure mode due to the concrete mix design, the conditions imposed on reports involving groundwater intrusion or concrete degradation, along with the structural health monitoring program, did not consider the possibility of ASR development.

One contributing cause was nonetheless identified: failure to prioritize groundwater elimination or mitigation resulted in a greater concrete area being exposed to moisture.

Material Degradation   Initial testing of extracted core samples indicated reductions in the modulus of elasticity values from those assumed in the original design. The first compressive strength tests from the electrical tunnel were compared to the original test cylinders cast during construction of the Control Building in 1979. This comparison also appeared initially to indicate an approximately 22% decrease in compressive strength. Extracted cores were expected to yield compressive strength values 10% to 15% lower than cylinder test results.

When additional cores were tested from both ASR-affected and non-ASR-affected areas, the tested compressive strengths were essentially the same, a finding consistent with the industry literature, which predicts minimal impact to tested compressive strength levels at relatively low ASR expansions. The modulus of elasticity equaled approximately 47% of the expected value (ML121160422, 2012; ML13151A328, 2013)

Structural Integrity and Testing In the most recent report ML13151A328 (2013), it was stated that the purpose of testing is to assess the impact of ASR on out-of-plane shear performance and reinforcement anchorage (lap splice) performance. Test specimens will use the walls in the B Electrical Tunnel as the reference location for the Seabrook Station, and the walls will be modeled as reinforced concrete beams constructed in order to be similar to the reference location walls. It is anticipated that testing will provide data to assess the effects of ASR on shear and reinforcement anchorage performance; whenever necessary, testing will assess the effectiveness of retrofit techniques in improving the structural capacity of beams at various levels of ASR degradation.

The following information will be developed by the Shear Test Program:

- Shear Capacity of ASR-affected Reinforced Concrete Beams: determines the extent to which shear performance of the reinforced concrete beams has been affected as a function of ASR degradation.
- Flexural Stiffness of ASR-affected Reinforced Concrete Beams: determines the extent to which flexural stiffness of the reinforced concrete beams has been affected as a function of ASR degradation.
- Efficacy of Retrofit Technique: determines the effectiveness of the given retrofit technique in enhancing shear performance as a function of ASR degradation. Test results may then be used to determine whether any margin exists between the actual (experimentally determined) shear strength of reinforced concrete beams and the calculated shear strength (by applying relevant provisions in the design code, ACI 318-71).

Regrettably, most figures and important information pertaining to these tests were blacked out[9]. The earlier report (ML121160349, 2012) on tests performed at the University of Texas at Austin must now be consulted in order to grasp the meaning of this test.

Test beams were fabricated to represent the Seabrook structural elements and had both varying levels of ASR and control beams with no ASR. The reinforcement detail depicts the lack of through-thickness reinforcement in tunnel walls and enables an in-depth study of shear and anchorage behavior at both the current and future levels of ASR degradation, Fig. 1.11. Preliminary results are shown in Figure 1.12(a), along with the application of test results in Figure 1.12(b).

---

[9]NextEra decided to withhold proprietary information from public disclosure as it alleged that the *information requested to be withheld represents the product of several years of intensive NextEra Energy Seabrook efforts and moreover a considerable expenditure. This information may be made available to the market in the event nuclear facilities or other regulated facilities identify the presence of ASR. For potential customers to duplicate this information, similar technical programs would have to be conducted, and significant staffing, with the requisite talent and experience, would have to be allocated. The extent to which this information is available to potential customers hinders NextEra Energy Seabrook's ability to sell products and services that use this information.*

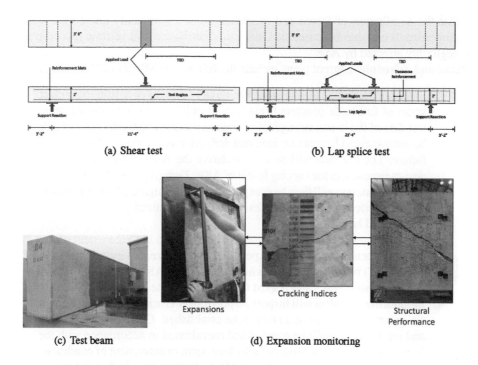

(a) Shear test   (b) Lap splice test

Cracking Indices

Expansions   Structural
                                                    Performance

(c) Test beam   (d) Expansion monitoring

FIGURE 1.11   Experimental tests at UT-Austin (ML121160349, 2012)

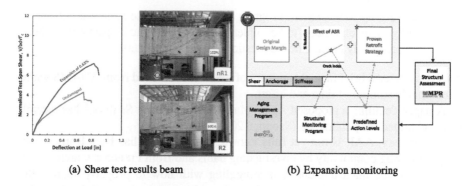

(a) Shear test results beam   (b) Expansion monitoring

FIGURE 1.12   Tests results, ML121160349 (2012)

Finite Element Studies A finite element model of the most limiting area was developed to address the potential of an adverse dynamic response associated with the apparent modulus of elasticity tests conducted on the extracted core samples. According to this model, a differential analysis of the structure with various modulus changes could be performed (ML13151A328, 2013). This analysis concluded that (ML121160422, 2012) the maximum acceleration profiles within the Structure Response Spectrum are not significantly affected by ASR properties and moreover that the distribution of forces and moments is not significantly altered by ASR properties. It can thus be concluded that load distribution and seismic response is negligibly affected by ASR.

Plans and Schedule Current plans include the following (ML121160422, 2012):

1. Shear and lap splice testing at the University of Texas at Austin to conduct series of full-scale concrete beam tests in order to provide representative test data of the in situ strength of restrained concrete elements. Beams will be instrumented to measure load and deflection at incremental steps up to failure. The test data will be used to derive the ASR impact on concrete design parameters for varying levels of ASR. Design parameters for ASR-affected concrete will then be compared to ACI Design Code requirements and reconciled with Seabrook design basis calculations.
2. Pullout and breakout tests.
3. Expansion testing of coarse aggregates to establish both the extent of aggregate reaction to date and expected additional reactivity/expansion in the future. The following tests will be performed: ASTM C 1260 - Mortar Bar Expansion Test - and ASTM C 1293 - Concrete Prism Test.
4. Monitoring: NextEra will inspect 20 previously crack indexed locations at six-month intervals until a rate can be established. Changes in crack size and crack indices will be trended and reevaluated in accordance with the Structural Monitoring Program. The long-term management of conditions via established crack index thresholds, indicating at which point action will be taken.
5. The Aging Management Program (AMP) for License Renewal will initially focus on criteria to be implemented for periodic inspections of the 20 previously crack indexed locations, at 6-month intervals.

This constituted a brief factual summary of what was retrieved from ADAMS within a reasonable amount of time and effort.

A paper, detailing a proposed Aging Management Plan for Seabrook has recently been submitted for publication (Saouma, 2013b).

In summary, AAR is not yet fully known, though much has been learned from the engineering community by investigating dams and from French researchers. Not surprisingly, the nuclear industry is struggling with how to properly address this issue, and this author's advice is to take a closer look at the State-of-the-Art rather than relying on mere speculation.

1.3 A BRIEF REVIEW OF FINITE ELEMENT

Nowadays, the numerical modeling of AAR is almost exclusively accomplished using the finite element method. A brief review of the essentials of this method is thus warranted. This section could be easily skipped by the casual reader.

1.3.1 ELEMENT FORMULATION

We define an arbitrary solid, of volume Ω and boundary surfaces $\Gamma = \Gamma_t \bigcup \Gamma_u$, where traction t is applied over Γ_t, and displacements imposed over Γ_u. The only other force would be b the body force. Ω may be subjected to initial strains ϵ_0 and stresses σ_0, and has a constitutive relation given by $\sigma = D\epsilon$ (Eq. 1.34). The functional of the general form of the total potential energy (TPE) of this solid will be given by

$$\Pi = \underbrace{\frac{1}{2}\int_\Omega \epsilon^T D\epsilon d\Omega - \int_\Omega \epsilon^T D\epsilon_0 d\Omega + \int_\Omega \epsilon^T \sigma_0 d\Omega}_{U}$$

$$\underbrace{- \int_\Omega u^T b d\Omega - \int_{\Gamma_t} u^T \hat{t} d\Gamma - uP}_{-W_e} \tag{1.5}$$

A variational statement is obtained by taking the first variation of Π and setting this scalar quantity equal to zero

$$\delta\Pi = \int_\Omega \delta\epsilon^T D\epsilon d\Omega - \int_\Omega \delta\epsilon^T D\epsilon_0 d\Omega$$

$$+ \int_\Omega \delta\epsilon^T \sigma_0 d\Omega - \int_\Omega \delta u^T b d\Omega - \int_{\Gamma_t} \delta u^T \hat{t} d\Gamma = 0 \tag{1.6}$$

yielding the *Principle of Virtual Work*. We can write the strain-displacement relations in terms of a linear differential operator L

$$\epsilon = Lu \tag{1.7}$$

where

$$L = \begin{bmatrix} \frac{\partial}{\partial x} & 0 & 0 \\ 0 & \frac{\partial}{\partial y} & 0 \\ 0 & 0 & \frac{\partial}{\partial z} \\ \frac{\partial}{\partial y} & \frac{\partial}{\partial x} & 0 \\ \frac{\partial}{\partial z} & 0 & \frac{\partial}{\partial x} \\ 0 & \frac{\partial}{\partial z} & \frac{\partial}{\partial y} \end{bmatrix} \tag{1.8}$$

Since the differential operator L is linear, the variation of the strains $\delta\epsilon$ can be expressed in terms of the variation of the displacements δu

$$\delta\epsilon = \delta(Lu) = L\delta u \tag{1.9}$$

The variational statement can now be written as:

$$\delta\Pi = \underbrace{\int_\Omega \delta(Lu)^T D(Lu)d\Omega - \int_\Omega \delta(Lu)^T D\epsilon_0 d\Omega + \int_\Omega \delta(Lu)^T \sigma_0 d\Omega}_{\delta U}$$

$$\underbrace{- \int_\Omega \delta u^T b d\Omega - \int_{\Gamma_t} \delta u^T \hat{t} d\Gamma}_{-\delta W_e} = 0 \tag{1.10}$$

This equation constitutes the root of the formulation of the finite element method (Zienkiewicz et al., 2005) through the following simple steps:

Discretization of Equation 1.10 will be performed on an element domain Ω_e. The first step in the discretization process consists of defining the displacements **u** at a point inside the element in terms of the shape functions **N** and nodal displacements $\bar{\mathbf{u}}_e$ for the element

$$\mathbf{u} \overset{\text{def}}{=} \mathbf{N}\bar{\mathbf{u}}_e \tag{1.11}$$

The virtual displacements $\delta\mathbf{u}$ at a point inside the element can also be defined in terms of the shape functions **N** and nodal virtual displacements $\delta\bar{\mathbf{u}}_e$ for the element

$$\delta\mathbf{u} = \mathbf{N}\delta\bar{\mathbf{u}}_e \tag{1.12}$$

since $\boldsymbol{\epsilon} = \mathbf{L}\mathbf{u} = \mathbf{L}\mathbf{N}\bar{\mathbf{u}}_e$; $\delta\boldsymbol{\epsilon} = \delta(\mathbf{L}\mathbf{u}) = \mathbf{L}\mathbf{N}\delta\bar{\mathbf{u}}_e$ and in defining the discrete strain-displacement operator **B** as

$$\mathbf{B} \overset{\text{def}}{=} \mathbf{L}\mathbf{N} \tag{1.13}$$

Substitution after discretization into Eq. 1.10:
Strain energy for an element

$$\int_{\Omega_e} \delta(\mathbf{L}\mathbf{u})^T \mathbf{D}(\mathbf{L}\mathbf{u}) d\Omega = \delta\bar{\mathbf{u}}_e^T \underbrace{\int_{\Omega_e} \mathbf{B}^T \mathbf{D}\mathbf{B} d\Omega}_{\mathbf{K}_e} \bar{\mathbf{u}}_e \tag{1.14}$$

which defines the element stiffness matrix $\mathbf{K}_e$.
Similarly, we discretize the component of the initial strain ϵ_0 and stress σ_0 in Eq. 1.10 through **B**

$$\int_{\Omega_e} \delta(\mathbf{L}\mathbf{u})^T \mathbf{D}\epsilon_0 d\Omega + \int_{\Omega_e} \delta(\mathbf{L}\mathbf{u})^T \sigma_0 d\Omega =$$
$$\delta\bar{\mathbf{u}}_e^T \underbrace{\left[\int_{\Omega_e} \mathbf{B}^T \mathbf{D}\epsilon_0 d\Omega + \delta\bar{\mathbf{u}}_e^T \int_{\Omega_e} \mathbf{B}^T \sigma_0 d\Omega \right]}_{\mathbf{f}_{0_e}} \tag{1.15}$$

which defines the initial force vector $\mathbf{f}_{0_e}$ as the strain energy due to initial strains and stresses becoming
External work: The volume integral defining the work done by body forces and the surface integral defining the work done by surface traction

$$\int_{\Omega_e} \delta\mathbf{u}^T \mathbf{b} d\Omega + \int_{\Gamma_t} \delta\mathbf{u}^T \hat{\mathbf{t}} d\Gamma = \delta\bar{\mathbf{u}}_e^T \underbrace{\left[\int_{\Omega_e} \mathbf{N}^T \mathbf{b} d\Omega + \int_{\Gamma_t} \mathbf{N}^T \hat{\mathbf{t}} d\Gamma \right]}_{\mathbf{f}_e} \tag{1.16}$$

which defines the applied force vector $\mathbf{f}_e$

Final Rearrangement Having obtained the discretization of the various integrals
defining the variational statement for the total potential energy variational principle, it is now possible to define the discrete system of equations. Substituting
Equations 1.14, 1.15 and 1.16 into Equation 1.10 and rearranging terms, the discretized Principle of Virtual Work reduces to

$$\delta \bar{\mathbf{u}}_e^T \mathbf{K}_e \bar{\mathbf{u}}_e = \delta \bar{\mathbf{u}}_e^T \mathbf{f}_e + \delta \bar{\mathbf{u}}_e^T \mathbf{f}_{0_e} \tag{1.17}$$

Since $\delta \bar{\mathbf{u}}_e^T$ is an arbitrary (i.e. non-zero) vector appearing on both sides of Equation
1.17, the discrete system of equations can be simplified into

$$\mathbf{K}_e \bar{\mathbf{u}}_e = \mathbf{f}_e + \mathbf{f}_{0_e} \tag{1.18}$$

as the discrete system of equations for an element.

1.3.2 ISOPARAMETRIC ELEMENTS

Isoparametric elements are by far the most widely used types of elements in finite
element analysis, and their stiffness matrix (Eq. 1.14) is evaluated through numerical
integration,

$$
\begin{aligned}
[\mathbf{k}] &= \int_{-1}^{1} \int_{-1}^{1} [\mathbf{B}(\eta,\xi)]^T [\mathbf{D}(\eta,\xi)][\mathbf{B}(\eta,\xi)]t|\mathbf{J}(\eta,\xi)| d\xi d\eta \\
&\approx \sum_{i=1}^{n} \sum_{j=1}^{m} W_i^{(m)} W_j^{(n)} F(\xi_i, \eta_j [\mathbf{B}(\eta,\xi)]^T [\mathbf{D}(\eta,\xi)][\mathbf{B}(\eta,\xi)]t|\mathbf{J}|
\end{aligned}
\tag{1.19}
$$

whereby the integral is replaced by a numerical evaluation at each of the nm integration (Gauss) points. At each point $((\eta, \xi)$,$\mathbf{D}(\eta, \xi)$ is the corresponding constitutive
model derived from the free energy, such as in Eq. 1.34, Figure 1.13. The total number
of integration points will thus be $m \times n$ (3×3 for 2D eight-node isoparametric elements). Finally, in the context of a nonlinear incremental analysis, we would express

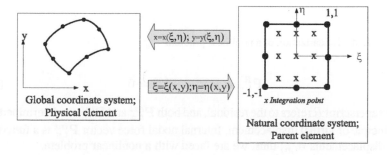

FIGURE 1.13   Gauss-Legendre Integration Over a Surface

the incremental stress $\dot{\sigma}$ in terms of the incremental strain $\dot{\epsilon}$ as

$$\dot{\sigma} = \dot{D}_T \dot{\epsilon} \qquad (1.20)$$

where $\dot{D}_T$ is the tangential material matrix (from which the tangential stiffness matrix will be derived.

1.3.3 NONLINEAR SYSTEM

Eq. 1.18 is in fact a statement of Newton's Third Law (When a first body exerts a force $\mathbf{P}_{ext}$ on a second body, the second body simultaneously exerts a force $\mathbf{P}_{int} = \mathbf{P}_{ext}$) or

$$\underbrace{\mathbf{f}_e + \mathbf{f}_{0_e}}_{\mathbf{P}_{ext}} - \underbrace{\mathbf{K}_e \overline{\mathbf{u}}_e}_{\mathbf{P}_{int}} = 0 \qquad (1.21)$$

whereas this can be exactly satisfied in a linear elastic system, it is only nearly satisfied in a nonlinear one, with the discrepancy being

$$\mathbf{P}^R = \mathbf{P}_{ext} - \mathbf{P}_{int} \qquad (1.22)$$

In the context of nonlinear analysis, $\mathbf{P}^{int} = \int \mathbf{B}^T \sigma d\Omega$. Generalizing, we can rewrite the preceding equation at time t, increment n (Figure 1.14) as:

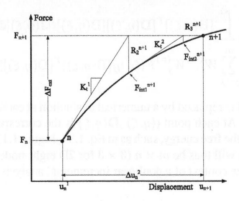

FIGURE 1.14   Linearization of a nonlinear problem

$$\mathbf{P}^R_{t,n} = \mathbf{P}^{ext}_{t,n} - \mathbf{P}^{int}_{t,n} = 0 \qquad (1.23)$$

where superscript R refers to the residual, and both $\mathbf{P}^{ext}_{t,n}$ and $\mathbf{P}^{int}_{t,n}$ are determined from the principle of virtual displacement. Internal nodal force vector $\mathbf{P}^{int}_{t,n}$ is a function of nodal displacements $\mathbf{u}_{t,n}$; thus, we are faced with a nonlinear problem.

Within each iteration, we determine the residual nodal force vector, yielding an incremental nodal displacement vector. Iterations continue until the residual nodal force vector or incremental nodal displacement vector is sufficiently small.

The problem is linearized (through a truncated Taylor series) into the following problem:

$$\mathbf{P}_{t,n}^{R,k} = \mathbf{P}_{t,n}^{ext} - \mathbf{P}_{t,n}^{int,k} \tag{1.24}$$

$$\delta\mathbf{u}_{t,n}^{k} = [\mathbf{K}_{tt,n}^{k-1}]^{-1} \cdot \mathbf{P}_{t,n}^{R,k} \tag{1.25}$$

$$\mathbf{u}_{t,n}^{k} = \mathbf{u}_{t,n}^{k-1} + \delta\mathbf{u}_{t,n}^{k} \tag{1.26}$$

$$\mathbf{u}_{t,n}^{k=0} = \mathbf{u}_{t,n-1} \tag{1.27}$$

$$\mathbf{P}_{t,n}^{int,k=0} = \mathbf{P}_{t,n-1}^{int} \tag{1.28}$$

where, subscript n refers to the load increment, and subscript k to the iteration number within a load increment. We thus assume equilibrium has been reached at increment n, at which point we apply an increment of external force $\Delta\mathbf{P}^{ext}$, and seek to determine the corresponding incremental displacement $\Delta\mathbf{u}_{n+1}$. The internal forces and corresponding displacements will then be in (near) equilibrium.

We distinguish between a load increment, and the iterations within an increment to reach equilibrium; at at each iteration, we can determine the residual $\mathbf{R}_i^{(n+1)}$ which corresponds to $\mathbf{P}_{ext} - \mathbf{P}_{int}$, and seek to minimize this residual. At each iteration, we update (according to the Newton method) the tangent stiffness matrix which corresponds to the Jacobian.

This modeling set-up is focused on determining the internal nodal force vector $\mathbf{P}_{t,n}^{int,k}$, and tangent stiffness matrix $\mathbf{K}_{tt,n}^{k-1}$.

Alternative integration schemes are shown in Fig. 1.15. In the Newton-Raphson method, the tangent stiffness matrix is updated at each iteration, in the Modified Newton-Raphson it is updated at each increment, and in the Initial stiffness method there is no update. The secant method is a numerical procedure to estimate the tangent stiffness matrix.

Not shown, is the arc-length method which should be used for softening material if one goes beyond peak load.

1.3.4 CONSTITUTIVE MODEL D

The Helmholtz free energy Ψ is a thermodynamic potential that describes the capacity of a system to do work from a closed thermodynamic system at constant temperature and volume (Malvern, 1969).

In order to maintain linear theory, it is sufficient to select as a convex thermodynamic potential a positive definite quadratic function in the components of the strain tensor:

$$\rho\Psi = \frac{1}{2\rho}\mathbf{a}:\boldsymbol{\epsilon}:\boldsymbol{\epsilon} \tag{1.29}$$

where $\rho\Psi$ is the specific energy and $\mathbf{a}$ is a fourth-order symmetric tensor. The stress is then defined as:

$$\sigma = \rho\frac{\partial\Psi}{\partial\boldsymbol{\epsilon}} = \mathbf{a}:\boldsymbol{\epsilon} \tag{1.30}$$

which is Hooke's Law.

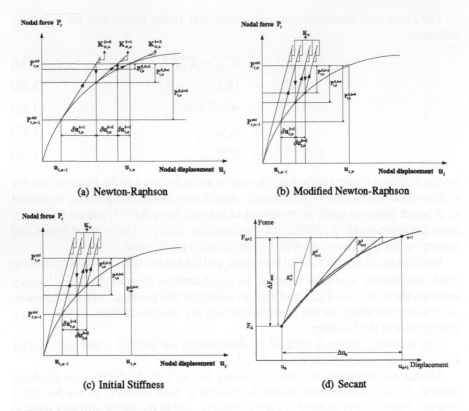

FIGURE 1.15   Integration schemes

Isotropy and linearity require that the potential Ψ be a quadratic invariant of the strain tensor, i.e. a linear combination of the square of the first invariant $\epsilon_I^2 = [\text{tr}\,(\epsilon)]^2$, and the second invariant $\epsilon_{II}^2 = \frac{1}{2}\text{tr}\,(\epsilon^2)$

$$\Psi = \frac{1}{2\rho}\left(\lambda\epsilon_I^2 + 4\mu\epsilon_{II}\right) \tag{1.31}$$

where λ and μ are Lame's parameters

$$\lambda = \frac{E\nu}{(1+\nu)(1-2\nu)}; \qquad \mu = \frac{E}{2(1+\nu)} \tag{1.32}$$

In differentiating we obtain the stress:

$$\sigma = \rho\frac{\partial\Psi}{\partial\epsilon} = \lambda\text{tr}\,(\epsilon)\mathbf{I} + 2\mu\epsilon \tag{1.33}$$

Expanding into 3D Cartesian coordinates this equation yields:

$$
\underbrace{\begin{Bmatrix} \sigma_{xx} \\ \sigma_{yy} \\ \sigma_{zz} \\ \tau_{xy} \\ \tau_{yz} \\ \tau_{zx} \end{Bmatrix}}_{\sigma} = \underbrace{\left[\begin{array}{cc} \frac{E}{(1+\nu)(1-2\nu)}\begin{bmatrix} 1-\nu & \nu & \nu \\ \nu & 1-\nu & \nu \\ \nu & \nu & 1-\nu \end{bmatrix} & 0 \\ 0 & G\begin{bmatrix} 1 & 0 & 0 \\ 0 & 1 & 0 \\ 0 & 0 & 1 \end{bmatrix} \end{array} \right]}_{D}
$$
$$
\underbrace{\left\lfloor \epsilon_{xx} \quad \epsilon_{yy} \quad \epsilon_{zz} \quad \gamma_{xy}(2\epsilon_{xy}) \quad \gamma_{yz}(2\epsilon_{yz}) \quad \gamma_{zx}(2\epsilon_{zx}) \right\rfloor^T}_{\epsilon} \quad (1.34)
$$

For problems involving a long body in the z direction with no variation in load or geometry (such as dams), then $\varepsilon_{zz} = \gamma_{yz} = \gamma_{xz} = \tau_{xz} = \tau_{yz} = 0$. Thus, Eq. 1.34 is simplified to

$$
\begin{Bmatrix} \sigma_{xx} \\ \sigma_{yy} \\ \sigma_{zz} \\ \tau_{xy} \end{Bmatrix} = \frac{E}{(1+\nu)(1-2\nu)} \underbrace{\begin{bmatrix} (1-\nu) & \nu & 0 \\ \nu & (1-\nu) & 0 \\ \nu & \nu & 0 \\ 0 & 0 & \frac{1-2\nu}{2} \end{bmatrix}}_{D} \begin{Bmatrix} \varepsilon_{xx} \\ \varepsilon_{yy} \\ \gamma_{xy} \end{Bmatrix} \quad (1.35)
$$

Finally, in plane stress conditions the longitudinal dimension in z direction is much smaller than in the x and y directions, then $\tau_{yz} = \tau_{xz} = \sigma_{zz} = \gamma_{xz} = \gamma_{yz} = 0$ throughout the thickness. Again, substituting into Eq. 1.34 we obtain:

$$
\begin{Bmatrix} \sigma_{xx} \\ \sigma_{yy} \\ \tau_{xy} \end{Bmatrix} = \frac{1}{1-\nu^2} \underbrace{\begin{bmatrix} 1 & \nu & 0 \\ \nu & 1 & 0 \\ 0 & 0 & \frac{1-\nu}{2} \end{bmatrix}}_{D} \begin{Bmatrix} \varepsilon_{xx} \\ \varepsilon_{yy} \\ \gamma_{xy} \end{Bmatrix} \quad (1.36\text{-a})
$$

$$
\varepsilon_{zz} = -\frac{1}{1-\nu}\nu(\varepsilon_{xx} + \varepsilon_{yy}) \quad (1.36\text{-b})
$$

Finally, in the presence of chemically induced deterioration, the free energy Ψ can be enriched by an additional term. This term may be arbitrarily selected as long as it satisfies the first and second laws of thermodynamics. Hence, one could write:

$$
\Psi = \underbrace{\frac{1}{2\rho}\left(\lambda\epsilon_I^2 + 4\mu\epsilon_{II}\right)}_{\text{Mechanical}} + \underbrace{g(\xi)}_{\text{Chemical}} \quad (1.37)
$$

where $g(\xi)$ corresponds to the free energy due to the chemical deterioration (such as AAR expansion).

This last equation should be contrasted with Eq. 1.51 where the AAR strain is treated as an initial strain independent of the constitutive model, whereas in here, the chemical component (AAR) is interlaced with the mechanical stress-strain relation resulting in a coupled system.

Finally, Fig. 1.16 summarizes the fundamental equations in solid mechanics.

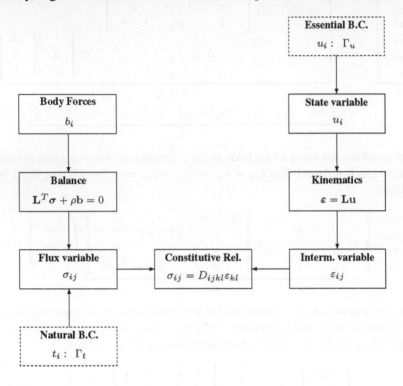

FIGURE 1.16   Fundamental equations in solid mechanics

1.4   A BRIEF REVIEW OF HEAT TRANSFER

AAR is very much dependent on temperature. Hence, a proper AAR numerical simulation should entail, amongst other, a proper transient heat transfer analysis to determine the time variation of temperature at each node.

1.4.1   MODES OF HEAT TRANSFER AND BOUNDARY CONDITIONS

There are three fundamental modes of heat transfer:

Conduction: is caused by microscopic diffusion and collisions of particles within a body as a result of a temperature gradient. It is governed by Fourier's Law

$$q_i^{cond} = -k_i \frac{\partial T}{\partial x_i} \tag{1.38}$$

where $T = T(x, y)$ is the temperature field in the medium, q_x and q_y are the components of the heat flux (W/m^2 or Btu/h.ft^2), k is the thermal conductivity (W/m.°C or Btu/h.ft.°F) and $\frac{\partial T}{\partial x}$, $\frac{\partial T}{\partial y}$ are the temperature gradients along the x and y respectively. The resultant heat flux $\mathbf{q} = q_x \mathbf{i} + q_y \mathbf{j}$ is at right angles to an isotherm or a line of constant temperature. The minus sign indicates that flux is along the direction of decreasing temperature.

Convection: is the transfer of heat form one point to another through the movement of fluid. Hence, it can only occur when a material is exposed to a moving fluid (such as air or water) which is at different temperature. It is governed by Newton's Law of Cooling

$$q^{conv} = h(T_s - T_\infty) \tag{1.39}$$

where q is the convective heat flux (W/m^2), h is the convection heat transfer coefficient or film coefficient (W/m^2.°C or Btu/h.ft^2.°F). It depends on various factors, such as whether convection is natural or forced, laminar or turbulent flow, type of fluid, and geometry of the body; T_s and T_∞ are the surface and fluid temperature, respectively.

Radiation: is the electromagnetic radiation generated by the thermal motion of charged particles in matter. All matter with a temperature greater than absolute zero emits thermal radiation. Thermal radiation is thus the energy transferred between two separated bodies at different. The fundamental law is the Stefan-Boltman's Law of Thermal Radiation for black bodies in which the flux is proportional to the fourth power of the absolute temperature which causes the problem to be nonlinear.

$$q^{rad} = eC_s \left(T_{surf}^4 - T_{fluid,\infty}^4 \right) \tag{1.40}$$

where e is the emissivity, C_s the Stefan-Boltzman's constant.

Heat exchange modes govern the possible boundary conditions in thermal problems.

Essential: Temperature prespecified on Γ_T $(T = T_0)$
Natural: Flux
  Specified Flux prescribed on Γ_q, $q_n = q_0$, note an insulated surface will have zero flux across it, thus $q_n = 0$.
  Convection Flux prescribed on Γ_c, $q_c = h(T - T_\infty)$. Note this type of boundary condition is analogous to the one in structural mechanics where we have an inclined support on rollers.

With reference to a dam, those boundary conditions are illustrated in Fig. 1.17.

1.4.2 GOVERNING PARTIAL DIFFERENTIAL EQUATION

If we consider a unit thickness, 2D differential body of dimensions dx by dy, Fig. 1.18, then the rate of heat generation/sink is

$$I_2 = Q \mathrm{d}x \mathrm{d}y \tag{1.41}$$

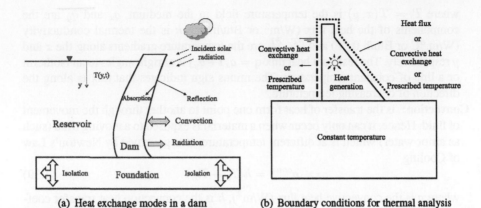

(a) Heat exchange modes in a dam   (b) Boundary conditions for thermal analysis

FIGURE 1.17   Thermal analysis of a dam

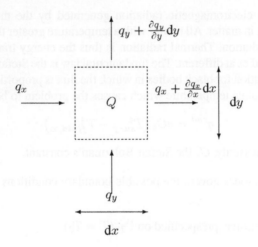

FIGURE 1.18   Flux through sides of a differential element

where Q (W/m^3) is the rate of heat (positive) or sink (negative) generation. Heat flux across the boundary of the element is given by (note similarity with equilibrium equation)

$$I_1 = \left[\left(q_x + \frac{\partial q_x}{\partial x} dx \right) - q_x dx \right] dy + \left[\left(q_y + \frac{\partial q_y}{\partial y} dy \right) - q_y dy \right] dx$$
$$= \frac{\partial q_x}{\partial x} dx dy + \frac{\partial q_y}{\partial y} dy dx$$

(1.42)

where q is the flux vector. The Change in stored energy is

$$I_3 = c\rho \frac{dT}{dt} . dxdy \qquad (1.43)$$

where c, the specific heat, is the amount of heat per unit mass to cause a unit rise in temperature, ρ is the mass density, T is the temperature,

From the first law of thermodynamics, energy produced I_2 plus the net energy across the boundary I_1 must be equal to the energy absorbed I_3, thus

$$I_1 + I_2 - I_3 \;=\; 0 \qquad (1.44\text{-a})$$

$$\underbrace{\frac{\partial q_x}{\partial x} dxdy + \frac{\partial q_y}{\partial y} dydx}_{I_1} + \underbrace{Qdxdy}_{I_2} - \underbrace{c\rho \frac{dT}{dt} dxdy}_{I_3} \;=\; 0 \qquad (1.44\text{-b})$$

This equation can be rewritten as

$$\frac{\partial q_x}{\partial x} + \frac{\partial q_y}{\partial y} + Q = \rho c \frac{\partial T}{\partial t} \qquad (1.45)$$

We note the similarity between this last equation (scalar), and the equation of equilibrium equation (vectorial)

$$\frac{\partial \sigma_{xx}}{\partial x} + \frac{\partial \sigma_{xy}}{\partial y} + \rho b_x \;=\; \rho m \frac{\partial^2 u_x}{\partial t^2} \qquad (1.46\text{-a})$$

$$\frac{\partial \sigma_{yy}}{\partial y} + \frac{\partial \sigma_{xy}}{\partial x} + \rho b_y \;=\; \rho m \frac{\partial^2 u_y}{\partial t^2} \qquad (1.46\text{-b})$$

For steady state problems, the previous equation does not depend on t, and for 2D problems, it reduces to

$$\left[\frac{\partial}{\partial x} \left(k_x \frac{\partial T}{\partial x} \right) + \frac{\partial}{\partial y} \left(k_y \frac{\partial T}{\partial y} \right) \right] + Q = 0 \qquad (1.47)$$

For steady state isotropic problems,

$$\frac{\partial^2 T}{\partial x^2} + \frac{\partial^2 T}{\partial y^2} + \frac{\partial^2 T}{\partial z^2} = -\frac{Q}{k} \qquad (1.48)$$

which is Poisson's equation in 3D. Finally, if the heat input $Q = 0$, then the previous equation reduces to

$$\frac{\partial^2 T}{\partial x^2} + \frac{\partial^2 T}{\partial y^2} + \frac{\partial^2 T}{\partial z^2} = 0 \qquad (1.49)$$

which is an Elliptic (or Laplace) equation. Solutions of Laplace equations are termed *harmonic functions* (right hand side is zero).

If the function depends only on x and t, then we obtain

$$\rho c \frac{\partial T}{\partial t} \;=\; \frac{\partial}{\partial x} \left(k_x \frac{\partial T}{\partial x} \right) + Q \qquad (1.50)$$

which is the classical parabolic (or Heat) equation. Finally, Fig. 1.19 summarizes the fundamental equations in heat flow.

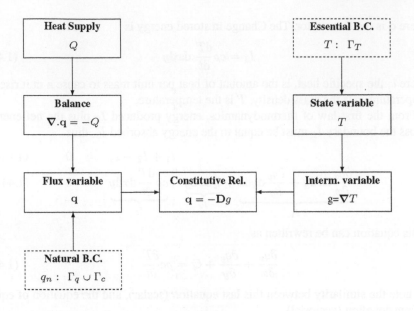

FIGURE 1.19   Fundamental equations of heat flow

1.5   FINITE ELEMENT MODELING OF AAR

1.5.1   SCALE AND MODELS

The modeling of AAR expansion has been undertaken by various researchers, (Pian et al., 2012). Generally speaking, this modeling effort falls into one of three categories:

Micro Models: A single aggregate is (typically) modelled along with the cement paste. The transport equation is used to model gel, whose formation is first accommodated by its saturation of the capillary pores, (Furusawa et al., 1994), (Bažant, Z.P and Steffens, A., 2000), (Lemarchand et al., 2001), (Suwito et al., 2002). While essential to properly understanding the underlying phenomenon causing AAR, this level of modeling, is of little relevance to a structural analysis of AAR-affected structures (i.e. the main topic of this book).

Meso Models An extension of the previous approach, in which both particle size distribution and interaction among individual particles are taken into account, (Comby-Peyrot et al., 2009) (Dunant and Scrivener, 2010).

Macro Models: This set-up avoids transport modeling, with emphasis placed on a comprehensive numerical model for the analysis of a given structure. Some (older) emperical models fully decouple structural modeling from reaction kinetics, while others couple these two effects (and may ignore the kinetics altogether).

   Empirical models were first presented by Charlwood et al. (1992) and Thompson et al. (1994) who identified critical issues related to AAR, namely stress dependency, i.e. no AAR expansion under a compressive stress of around 8 MPa and an expansion akin to thermal expansion. Subsequently, more refined models were presented by Léger et al. (1996), Bournazel and Moranville (1997), Capra and Bournazel (1998), and Herrador et al. (2009). In this context, the work of Huang and Pietruszczak (1999) stands apart since it may be one of the earliest contributions in which the kinetics are taken into account in terms of both temperature and confining stresses. Publication of Larive (1998b)'s work brought an end to the "era" of the parametric model and opened the door to developing what came to be known as "coupled chemo-mechanical" based models.

Coupled Chemo-Mechanical are models which account (indirectly) for the time dependency of the (chemical) diffusion in the mechanical response. For the most part, these models are inspired from and calibrated with the experiments of Larive (1998b) and Multon et al. (2004). They will be briefly reviewed in the following section.

Some current research work is seeking to develop a "grand unifying" multi-scale model that bridges the gap from micro to meso to macro in order to better capture the reaction (at the micro-scale) and its implications (on the macro scale). Puatatsananon and Saouma (2013) is one such attempt to bridge the gap between the micro and meso levels.

1.5.2 OVERVIEW OF COUPLED CHEMO-MECHANICAL MODELS

Broadly speaking, there are four "schools", defined as institutions, that have produced two or more publications focusing on coupled chemo-mechanical models. For the most part, these publications offer refinements of an original model.

1. The work by *Laboratoire Central des Ponts et Chaussées* (LCPC) can be traced back to the research of Ulm et al. (2000) followed by Li and Coussy (2002) and later Seignol (2011). The "LCPC" model was also refined by Farage et al. (2004) and Fairbairn et al. (2006). In most cases these models are rooted in the theory of plasticity and implemented within the LCPC finite element code CESAR.
2. Toulouse and EdF have had a longstanding collaboration that has given rise to what could be called the "EdF" model. As opposed to the LCPC model, the Toulouse/EdF developments are rooted in Damage Mechanics pioneered at the ENS/Cachan where Sellier performed his original work, (Sellier et al., 1995). Following a refined model for the influence of water (Poyet et al., 2006), Grimal et al. (2008) developed a visco-elasto-plastic orthotropic damage model which was further refined in (Sellier et al., 2009) which is described in some details in Sect. 9.1.3. These models are mainly based on Damage Mechanics (as opposed to plasticity) and implemented in EdF's computer programming Code ASTER.
3. The *Politecnico di Milano* has also developed coupled chemo-mechanical models, similar those of Toulouse/EdF, (Comi and Perego, 2011) (Comi et al., 2012); they have been used to analyze some complex problems.

4. Finally, Colorado has developed a coupled model which will be described in the next chapter.
5. Others important models include the one of Bangert et al. (2004)

Broadly speaking, we consider two forms of coupling:

Loose Coupling Revisiting Eq. 1.18 applied to an AAR analysis in the absence of an initial stress ($\sigma_0 = 0$) yet with an AAR induced initial strain, we have $\epsilon_0 = \epsilon_{AAR}$ where ϵ_{AAR} is the AAR strain tensor which will be defined later by Eq. 2.1. Hence, Eq. 1.18 becomes

$$\underbrace{\int_{\Omega_e} \mathbf{B}^T \mathbf{D} \mathbf{B} d\Omega \, \overline{\mathbf{u}}_e}_{\mathbf{K}_e} = \underbrace{\int_{\Omega_e} \mathbf{N}^T \mathbf{b} d\Omega + \int_{\Gamma_t} \mathbf{N}^T \hat{\mathbf{t}} d\Gamma}_{\mathbf{f}_e} + \underbrace{\int_{\Omega_e} \mathbf{B}^T \mathbf{D} \epsilon_{AAR} d\Omega}_{\mathbf{f}_{AAR}} \quad (1.51)$$

and the AAR strain is completely uncoupled from the constitutive model $\mathbf{D}$.

Tight Coupling Rooted in Eq. 1.37 where the inclusion of the chemical potential (which accounts for AAR) results in a fully coupled constitutive matrix without explicit expression for the AAR strains yielding

$$\underbrace{\int_{\Omega_e} \mathbf{B}^T \mathbf{D}' \mathbf{B} d\Omega \, \overline{\mathbf{u}}_e}_{\mathbf{K}_e} = \underbrace{\int_{\Omega_e} \mathbf{N}^T \mathbf{b} d\Omega + \int_{\Gamma_t} \mathbf{N}^T \hat{\mathbf{t}} d\Gamma}_{\mathbf{f}_e} \quad (1.52)$$

where $\mathbf{D}'$ is the constitutive matrix which embodies both the mechanical stress strain relation and the effects of the chemical reaction (Eq. 1.37), Fig. 1.20.

Henceforth, except for the Colorado model (the subject of this book), all other models rely on a single constitutive model that couples chemical (AAR expansion) and mechanical (non-linear stress strain) components. In these models,

1. One does not get only an AAR model, but also an accompanying stress-strain nonlinear constitutive model to which it is grafted. This is a clear departure from earlier models where AAR strain was accounted for as an initial strain ϵ_0 which would find its place in Eq. 1.15 through $\mathbf{f}_{0_e}$.
2. In lieu of extracting AAR strains separately, one is limited to the AAR extent (ξ) or damage.

It can indeed be argued that there is no reason a priori why AAR expansion should alter an element stiffness matrix (except for the deterioration of E), and inserting AAR within $\mathbf{D}$ is an unnecessary complication (albeit one that leads to a very "elegant" formulation). From a conceptual perspective, this mirrors the ongoing discussion on crack modeling: smeared vs. discrete.

To the best of the author's knowledge, the only AAR model that can be easily associated with any existing constitutive model (even linear elastic) is the one proposed in Saouma and Perotti (2006). Among known implementations, are Pan et al. (2013); Mirzabozorg (2013). It is this model which will be at the core of this book.

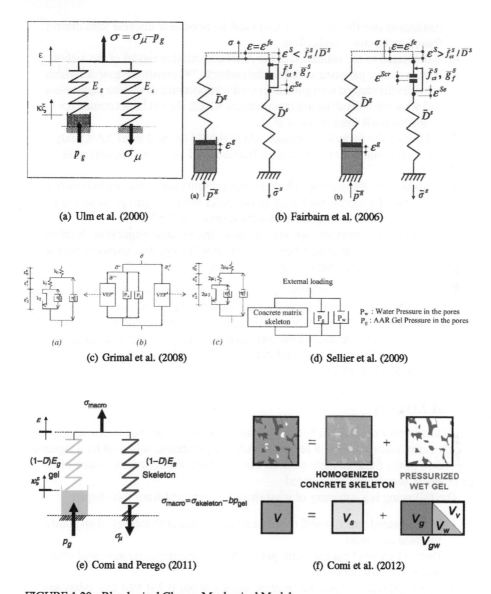

(a) Ulm et al. (2000)   (b) Fairbairn et al. (2006)

(c) Grimal et al. (2008)   (d) Sellier et al. (2009)

(e) Comi and Perego (2011)   (f) Comi et al. (2012)

FIGURE 1.20   Rheological Chemo-Mechanical Models

1.6   BOOK CONTENT

Following this first introductory chapter, the remaining ones

2. Details the AAR constitutive model developed by the authors in great details.
3. Begins by addressing the issue of concrete cracking (a natural result of AAR), and then presents two constitutive models for concrete: a) a nonlinear

continuum one (based on plasticity) and an nonlinear discrete joint elment (based on fracture mechanics).

4. is dedicated to the validation of the AAR constitutive model of the author.
5. Is an extensive parametric study of the authorâĿ™s constitutive model which provides the analyst with guidelines on the importance of various parameters and what are indeed the important effects which should be accounted for in a modern AAR finite element study.
6. Addresses the material properties which should be used in an AAR study.
7. Presents a number of different applications, all based on the authorsâĿ™ model.
8. Departs from the previous chapters and presents micro-model to simulate the AAR in a mortar bar based on diffusion models. Also presented is an innovative mathematical model for the kinetics of AAR.
9. Attempts to address the important issue of the residual expansion. In other words, how to determine when the reaction will stop. An innovative partial mathematical based model is also presented.

The first appendix presents numerical benchmarks problems which could be used by others to validate their finite element codes. Solution to these problems was presented in chapter 4. The second is a very brief description of the finite element code MERLIN developed by the author and which was used for the AAR analyses presented in chapter 7. Finally, the third one is a brief overview of chemical reaction kinetics to better comprehend chapter 8

1.7 SUMMARY

This introductory chapter has laid the groundwork for a proper understanding of the book's entire contents. First a review of AAR is presented, followed by a rigorous derivation of the finite element method. Finally, the finite elment modeling of AAR by others is very briefly categorized.

The following is a summary of what should be retained from this chapter:

1. Well vibrated concrete will exhibit a greater volumetric increase than poorly vibrated concrete!
2. Due to localized confinement, gel formation and pressurization will result in a volumetric increase.
3. The maximum volumetric increase equals about 0.2-0.5%; this value appears to be independent of temperature. The maximum volumetric strain is equivalent to a thermal expansion with a temperature increase of 200-500°C ($\alpha = 10^{-5}$).
4. The reaction kinetics follow an S-shaped curve: very slow at first, then increasing rapidly, and finally gradually slowing down again. As opposed to the maximum AAR-induced strain, the reaction rate is temperature-dependent.
5. A reduction in both tensile strength and elastic modulus occurs, in association with the microcracking caused by AAR.

6. Modern finite element codes for AAR modeling must account for chemo-mechanical based kinetics and fall into one of two categories: the AAR model being tightly coupled with the (mechanical) constitutive model; and the AAR being modeled with initial strain independence of the mechanical constitutive model.

7. Whereas dams have provided a major stimulus to investigate AAR, the presence of the later in NPP is a major challenge to the engineering community.

2 AAR Constitutive Model

This chapter will focus on the development of *a* constitutive model for the numerical modeling of chemically-induced expansive reactions (AAR, ASR and others). This model is developed at the macro-scale, does not depend on the concrete mix design and, as will be shown later, is uniquely characterized by three parameters. The relationship between micro (where diffusion processes are taken into account) and macro models will be addressed separately in Chapter 8.

It will be shown that, as opposed to most other existing models (except the over-simplified ones), this particular model is distinct from the intrinsic stress-strain constitutive model of concrete and hence could be easily grafted onto any existing linear/nonlinear finite element code.

2.1 MINIMUM REQUIREMENTS FOR A "MODERN" AAR NUMERICAL MODEL

Practically any model with sufficient "cursors" can be calibrated to yield what appears to be an acceptable correlation with field-measured displacements (and at times, which implies unrealistic values for some physical parameters). This step does not necessarily make the model "correct". Before a model can be considered adequate (in the sense of being able to generate long-term predictions), certain minimum requirements must first be met:

1. Spatial and temporal distribution of the structure temperature $T(x, y, z, t)$ and relative humidity $RH(x, y, z, t)$
2. Constitutive models (D) that can accommodate both a linear elastic response (for fast 3D analysis of the entire structure) and nonlinear response to account for cracking and failure.
3. Ability to properly model crack/joints. Vertical expansion is likely to cause either abutment cracking or "lift-off" of the concrete along (V-shaped) abutments or even within the inner center of rock-concrete interfaces (effect compounded by uplift). This is also required to capture the eventual closing of a sliced portion of the structure (as is often performed in dams to relieve stresses).
4. Creep to account for long-term deformation that can reduce AAR expansion.
5. Stress-induced anisotropy: confining stress will reduce expansion in the corresponding direction, but may increase it in orthogonal directions.
6. Ability to display AAR-induced strains (as opposed to the extent of AAR penetration (ξ).
7. Should seismic excitations be of concern, then the ability to perform a restart with the existing state of stress and internal degradation of $f'_t(x, y, z, t)$ and $E(x, y, z, t)$

To assess AAR-equipped finite element codes, a number of benchmark problems have been proposed. Regrettably, in most or all of these problems, the analyst must perform a comprehensive analysis of a complex structure (such as a dam), which makes it practically impossible to compare codes and pinpoint deficiencies. Accordingly, both the author and Sellier have proposed a battery of simple validation problems (of increasing complexity) in order to precisely ensure that a given code possesses all the necessary features for proper AAR modeling. This document is found in Appendix A.

Before any finite element code can be used for AAR studies, it must therefore be verified whether some or all of the features listed above are present; moreover, it would be preferable to determine how effectively the code is able to predict in the proposed battery of benchmark problems.

2.2 THE MODEL

A macro-based constitutive model for concrete expansion should account for: a) kinetics (i.e. time dependency) of the chemical reactions (due to ion diffusion); and b) the mechanics of the representative volume (e.g. effects of cracking and triaxial stress, property degradation).

Henceforth, the proposed model, (Saouma and Perotti, 2004) is based on the following considerations:

1. AAR model is completely separate from the constitutive model (linear or nonlinear stress-strain relations).
2. AAR is a volumetric expansion and, as such, cannot be addressed unidirectionally without due regard to what may occur along the other two orthogonal directions.
3. The expansion reaction is considered to be thermodynamically driven (i.e. time-dependent and affected by temperature) and has largely been inspired by the works of Larive (1998b) and Ulm et al. (2000).
4. AAR expansion is constrained by compression (Multon, 2004), and will be assumed redirected in other less constrained principal directions. This redirection will be accomplished through "weights" assigned to each of the three principal directions.
5. Relatively high compressive or tensile stresses inhibit AAR expansion due to the formation of micro (Hsu et al., 1963) or macro cracks, which absorb the expanding gel.
6. High compressive hydrostatic stresses slow the reaction, and a triaxial compressive state of stress reduces, without eliminating, expansion.
7. Accompanying AAR expansion entails a reduction in tensile strength and elastic modulus.

Based on these assumptions, the following general (uncoupled) equation for the incremental free volumetric AAR strain is given by:

$$\dot{\epsilon}_V^{AAR}(t,T) = \underbrace{\Gamma_t(f_t'|w_c, \sigma_I|COD_{max})\Gamma_c(\overline{\sigma}, f_c')}_{\text{Retardation}} \underbrace{g(h)}_{\text{Humidity}} \underbrace{\dot{\xi}(t,T)}_{\text{kinetics}} \underbrace{\epsilon(\infty)}_{\text{AAR Strain}}$$

(2.1)

Each of these terms will be addressed separately.

2.3 KINETICS

One of the most extensive and rigorous AAR investigations has been conducted by Larive (1998b), who tested more than 600 specimens with various mixes, ambient and mechanical conditions, and proposed a numerical model that governs concrete expansion. This thermodynamically-based, semi-analytical model was then calibrated using laboratory results in order to determine two key parameters: the latency time and characteristic times shown in Fig. 2.1(a) for the normalized expansion.

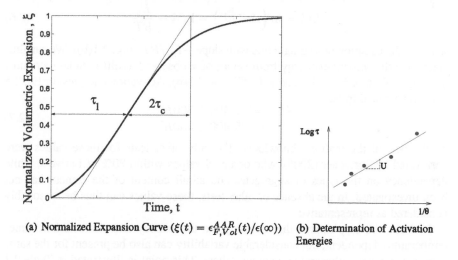

(a) Normalized Expansion Curve ($\xi(t) = \epsilon_{F,Vol}^{AAR}(t)/\epsilon(\infty)$)   (b) Determination of Activation Energies

FIGURE 2.1   Definition of the expansion curve

$$\xi(t,T) = \frac{1 - e^{-\frac{t}{\tau_c(T)}}}{1 + e^{-\frac{(t-\tau_l(T))}{\tau_c(T)}}}$$

(2.2)

or in rate form

$$\dot{\xi}(t,T) = \frac{e^{t/\tau_c}\left(e^{\frac{\tau_l}{\tau_c}} + 1\right)}{\tau_c\left(e^{t/\tau_c} + e^{\frac{\tau_l}{\tau_c}}\right)^2}$$

(2.3)

where τ_l and τ_c are the latency and characteristic times, respectively. The first corresponds to the inflection point, while the second is defined relative to the intersection of the tangent at τ_L with the asymptotic unit value of ξ. Like all chemical reactions, AAR is subject to Arrhenius Law (Arrhenius, 1889), which relates the dependence of the rate constant k of a chemical reaction on absolute temperature (T expressed in degrees Kelvin, $T^\circ K = 273 + T^\circ C$) and activation energy E_a.

$$k = Ae^{-\frac{E_a}{RT}} \tag{2.4}$$

substituting k with τ_L and τ_C, Ulm et al. (2000) has shown that these values at temperature T can be expressed in terms of the corresponding values at temperature T_0 through:

$$\begin{aligned} \tau_l(T) &= \tau_l(T_0) \exp\left[U_l\left(\frac{1}{T} - \frac{1}{T_0}\right)\right] \\ \tau_c(T) &= \tau_c(T_0) \exp\left[U_c\left(\frac{1}{T} - \frac{1}{T_0}\right)\right] \end{aligned} \tag{2.5}$$

where U_l and U_c are the activation energies required to trigger the reaction for latency and characteristic times, respectively. Activation energies can be easily determined by rewriting Eq. 2.5 in its non-exponential form:

$$\ln k = \ln\left(Ae^{-\frac{E_a}{RT}}\right) = \ln A - \frac{E_a}{RT} \tag{2.6}$$

which is the equation of a straight line with slope $-E_a/RT$ (Fig. 2.1(b)). We can thus determine the activation energy from values of k observed at different temperatures by simply plotting k as a function of $1/T$, Fig. 2.1(b). Activation energies for Eq. 2.5 were determined to be:

$$\begin{aligned} U_l &= 9{,}400 \pm 500K \\ U_c &= 5{,}400 \pm 500K \end{aligned} \tag{2.7}$$

To the best of the authors' knowledge, the only other tests for these values were performed by Scrivener (2005), who obtained values within 20% of Larive's, while dependency on the types of aggregates and alkali content of the cement has not been investigated. In the absence of other tests, these values can thus be reasonably considered as representative.

It should be emphasized that not only are the latency and characteristic times temperature-dependent, but considerable variability can also be present for the same concrete specimen chosen from among others. This point is illustrated in Table 2.1 for four specimens (ϕ13H24 kept at $38^\circ C$) tested by Larive (1998b).

2.3.1 SENSITIVITY TO TEMPERATURE

From Eq. 2.5, it is obvious that the reaction kinetics are highly dependent on temperature, which actually explains why some dams at high elevations throughout the world have exhibited AAR-induced expansion later when compared with those at lower elevations.

This temperature dependence is highlighted in Fig. 2.2(a), which illustrates the normalized expansion at four different temperatures. It is apparent that a major discrepancy exists in the expansion rate between 8° and 18°C, and then a smaller discrepancy between 18°, 28° and 38°C.

TABLE 2.1

Variation of $\epsilon(\infty)$, τ_c and τ_l for 4 specimens, (Larive, 1998b)

specimen		501	475	287	19	Mean	NSD (%)
$\epsilon(\infty)$	%	0.198	0.195	0.168	0.230	0.198	12.8
τ_c	days	19.9	35.3	25.8	22.0	25.7	26.5
τ_l	days	102.1	83.9	94.8	64.8	86.4	18.8
τ_l/τ_c	-	5.1	2.4	3.7	2.9	3.4	0.7

Fig. 2.2(b) illustrates the role of so-called "accelerated tests", many of which are performed at $38^\circ C$, in inducing expansion, whereas at ambient air temperature (in this case $7^\circ C$), expansion will be extremely slow. Then, Fig. 2.2(b) illustrates the

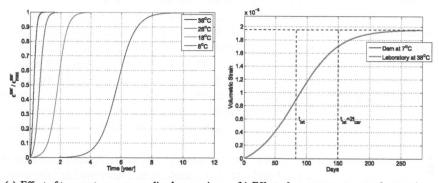

(a) Effect of temperature on normalized expansion   (b) Effect of temperature on actual expansion

FIGURE 2.2   Effect of Temperature on AAR Expansion

impact of temperature on both the latency and characteristic times.

2.3.2   SENSITIVITY TO INTEGRATION SCHEME

The time integration scheme may determine the expansion (in terms of temperature) through a secant approach ($\xi(t, T)$ from Eq. 8.37) or an incremental one ($\dot{\xi}(t, T)$ from Eq. 2.3. For constant or monotonically increasing temperatures, the two approaches would yied the same result. However, in the case where there may be a decrease in temperature, then erroneous results may be obtained. This will be illustrated by two simples examples. In the first, a harmonic variation of temperature is assumed over a year, and in the second an actual temperature record will be used.

For the harmonic temperature, one can assume

$$T_a = A \sin\left(\frac{2\pi(t-\xi)}{365}\right) + T_{mean} \qquad (2.8)$$

where $A = 0.5\left(|T_{max} - T_{mean}| + |T_{min} - T_{mean}|\right)$, t is the time in day, ξ is the time in day at which $T_a = T_{mean}$. Fig. 2.3(a) illustrates the temperature variation for T_{min}=14.2°C and T_{max}=16.2°C. The small difference is justified by the thermal inertia inside a dam. The mean and standard deviation f the concrete were selected to correspond to those assumed for a concrete dam in Coimbra (Portugal)[1] as illustrated below.

Assuming $t_{car} = 33.4$ days, $t_{lat} = 82.9$ days, U_{lat}=9400, U_{car}=5400, for a base temperature $T_0 = 48$°C, the corresponding yearly variations of the latency and characteristics times are shown in Figures 2.3(b) and 2.3(c) respectively for both assumptions: constant mean temperature of 15.2°C and variable one (oscillating between T_{min} and T_{max}). The higher values (thus slower reaction) is expected since the temperature is substantially lower than the reference one of 48°C.

When a constant mean temperature is used, both the secant (Eq. 8.37) and incremental (Eq. 2.3) approaches yield exactly the same expansion with time, Figure 2.3(d). However, both approaches differ drastically for the variable temperature shown in Fig. 2.3(a), as illustrated in Fig. 2.3(e) (where expansion is plotted with respect tot he right y axis, and temperature with respect to the left y axis). Finally, the incremental expansion first increases and then decreases smoothly in the case of a mean temperature, and irregularly if the time variation of temperature is accounted for.

Figure 2.3(f) compares variation of incremental strain for both mean and harmonic temperatures.

The same calculations are repeated for an actual dam location. Temperature record for Coimbra (Portugal) was used, and, Fig. 2.4(a) and it was determined that the mean air temperature is 15.2°C with a standard deviation of 4.9 °C. Due to the concrete thermal inertia, its temperature was estimated to have the same mean as the air, but a standard deviation of only 0.98°C.

As before, the temporal variations of both latency and characteristic times are shown in Figures 2.4(b) and 2.4(c) respectively first. Then Figures 2.4(d) and 2.4(e) compare AAR curve using mean and real temperatures. Finally, Figure 2.4(f) compares variation of incremental strain for both mean and harmonic temperatures.

From these figures, attention is drawn on the importance of using an incremental formulation for the correct AAR based expansion.

2.3.3 SENSITIVITY TO ACTIVATION ENERGIES

Given the limited laboratory data for activation energies, Fig. 2.5 shows the dependency of characteristic and latency times on a variation of their corresponding activation energies, in accordance with Eq. 2.7.

---

[1] Data was retrieved from National Oceanic and Atmospheric Administration (2013).

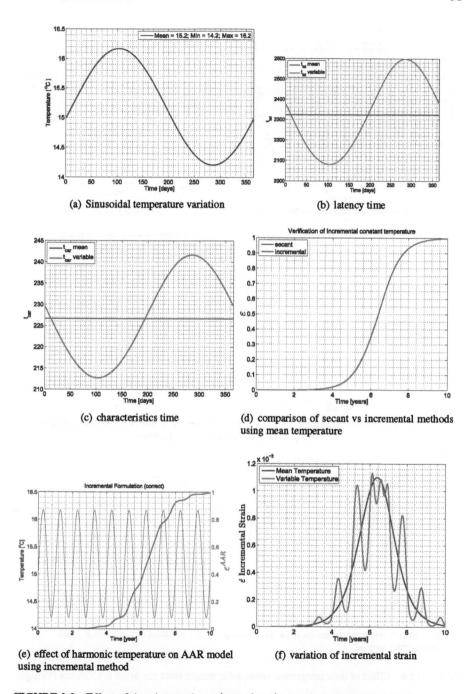

(a) Sinusoidal temperature variation

(b) latency time

(c) characteristics time

(d) comparison of secant vs incremental methods using mean temperature

(e) effect of harmonic temperature on AAR model using incremental method

(f) variation of incremental strain

FIGURE 2.3 Effect of time integration scheme for a harmonic temperature variation

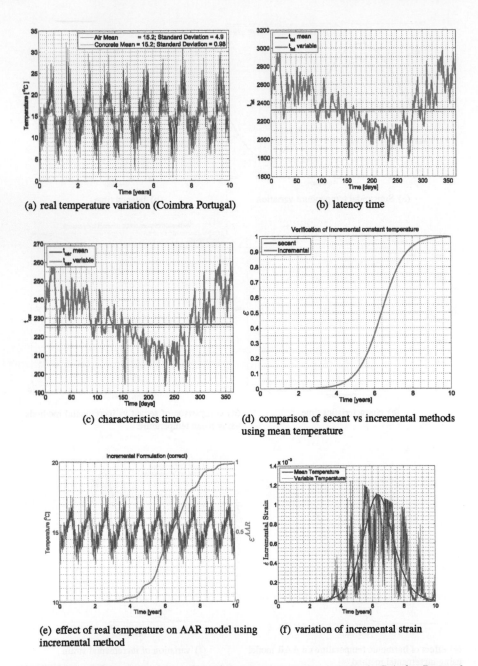

(a) real temperature variation (Coimbra Portugal)

(b) latency time

(c) characteristics time

(d) comparison of secant vs incremental methods using mean temperature

(e) effect of real temperature on AAR model using incremental method

(f) variation of incremental strain

FIGURE 2.4   Effect of time integration scheme for temperature variation in Coimbra Portugal

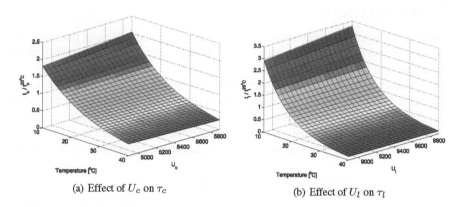

(a) Effect of U_c on τ_c (b) Effect of U_l on τ_l

FIGURE 2.5   Effects of activation energy variations on characteristic and latency times

2.3.4   SENSITIVITY TO TIME

The effect of time is clearly displayed in Fig. 2.6. In assuming values of $\epsilon(\infty)$ equal to 1, 10, and 100, with characteristic and latency times also increasing by a factor of 10, Fig. 2.6(a) shows that for all practical purposes, a linear increase in expansion over time may occur with no indication of "softening". This phenomenon has indeed been observed in certain dams, and Eq. 8.37 remains valid.

On the other hand, Fig. 2.6(b) illustrates the interdependency of time and temperature on the normalized expansion. It is clearly indicated that as temperature decreases, the expansion slows.

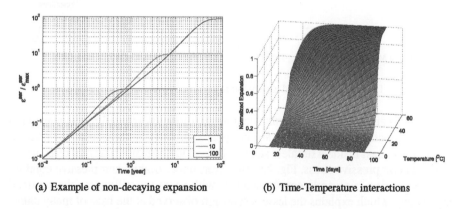

(a) Example of non-decaying expansion   (b) Time-Temperature interactions

FIGURE 2.6   Time Evolution

2.4 RETARDATION

A number of factors (besides temperatures) can slow the expansion of reactive concrete, namely: a) triaxial compressive stress, and b) presence of cracks. Each of these two factors will be examined separately.

By analogy with fatigue crack propagation, whereby an overstress may cause retardation of fatigue crack propagation (due to the presence of a plastic zone at the tip of the crack), the presence of micro or macro cracks will slow the expansion due to AAR. This observation is simply due to the fact that the gel generated would first have to fill the cracks before it could exert sufficient pressure to induce further expansion.

2.4.1 HYDROSTATIC COMPRESSIVE STRESS

It has long been established that a compressive stress (i.e. greater than about 8 MPa) will either limit or entirely prevent expansion in the corresponding direction. This finding was first stated by Hayward et al. (1988), as inspired by the relation between swelling pressures and swelling strains (for rocks inside tunnels) developed by Grob (1972) and Wittke and M. (2005). Multon (2004) addressed the issue of multi-axial

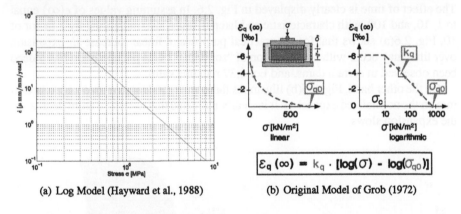

(a) Log Model (Hayward et al., 1988)   (b) Original Model of Grob (1972)

FIGURE 2.7   Uniaxially-restrained expansion model

confinement and showed that confinement in one direction will redirect expansion in the other directions (thus inducing anisotropic expansion). In these tests, the reactive concrete was cast inside a stainless steel cylinder and expansion was measured under a vertical compressive stress, Fig. 2.8. However, under triaxial compressive confinement, expansion is severely constrained (i.e. with room for redirecting the confined expansion), which explains the lesser expansion observed at the base of many dams, where the triaxial state of stress can be prevalent, as opposed to greater expansion in the upper part. This expansion constraint also explains why the unconstrained vertical expansion of the crest of a dam often provides a good indicator of the overall structural response for a finite element calibration, accounted for most conveniently by altering

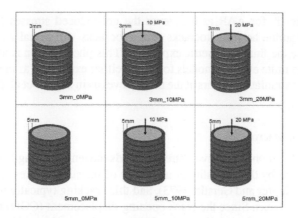

FIGURE 2.8 Tests on reactive concrete subjected to compressive stresses and confinement by Multon (Comi et al., 2009)

the latency time τ_l. Hence, Eq. 2.5 can be expanded as follows:

$$\tau_l(T, T_0, I_\sigma, f'_c) = f(I_\sigma, f'_c)\tau_l(T_0) \exp\left[U_l\left(\frac{1}{T} - \frac{1}{T_0}\right)\right] \tag{2.9}$$

where:

$$f(I_\sigma, f'_c) = \begin{cases} 1 & \text{if } I_\sigma \geqslant 0. \\ 1 + \alpha\frac{I_\sigma}{3f'_c} & \text{if } I_\sigma < 0. \end{cases} \tag{2.10}$$

and I_σ is the first invariant of the stress tensor($\frac{1}{3}(\sigma_I + \sigma_{II} + \sigma_{III})$), and f'_c the compressive strength.

Based on a careful analysis of Multon (2004), it was determined that $\alpha = 4/3$.

2.4.2 ROLE OF CRACKING

As a consequence of the AAR reaction, a gel is produced, Eq. 1.3, and this gel will first fill the adjoining pores, interfacial transition zone around the aggregates (caused by shrinkage of the cement paste) and the cracks, Fig. 2.9. Hence, the presence of cracks

FIGURE 2.9 Stress-Induced Cracks with Potential Gel Absorption, (Scrivener, 2003)

proves "beneficial" for AAR as it relieves the gel-induced stresses (at first). This section will examine how macrocracks or microcracks will actually absorb the gel and thus reduce the final volumetric expansion. This phenomenon is also embedded in most modern finite element models for AAR, either explicitly (as in this model) or explicitly when simulation occurs at the micro-level as in Suwito et al. (2002), Sellier et al. (2009).

2.4.2.1 Tensile Macrocrack

Tensile cracks occur once the tensile stress exceeds the tensile strength. Tensile cracking is not affected by the biaxiality or triaxiality of the state of stress (as opposed to compression) (Kupfer and Gerstle, 1973), and this cracking typically results in well-localized macrocracks. Since the crack opening displacement (COD) to account for gel absorption is needed, a nonlinear simulation could be performed in order to determine the COD. Such a step can be readily determined from so-called "discrete crack" models or, with more difficulty, when the smeared crack model is adopted.

This step will be accounted for in Eq. 2.1 through $\Gamma_t(f'_t|w_c, \sigma_I|COD_{max})$

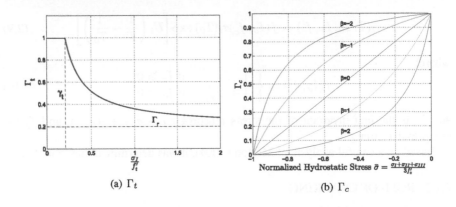

(a) Γ_t (b) Γ_c

FIGURE 2.10   Graphical representation of normalized retardation parameters Γ_c and Γ_t

$$\text{Smeared Crack} \begin{cases} \text{No } \Gamma_t = \begin{cases} 1 & \text{if } \sigma_I \leqslant \gamma_t f'_t \\ \Gamma_r + (1 - \Gamma_r)\gamma_t \frac{f'_t}{\sigma_I} & \text{if } \gamma_t f'_t < \sigma_I \end{cases} \\ \text{Yes } \Gamma_t = \begin{cases} 1 & \text{if } \text{COD}_{max} \leqslant \gamma_t w_c \\ \Gamma_r + (1 - \Gamma_r)\gamma_t \frac{w_c}{\text{COD}_{max}} & \text{if } \gamma_t w_c < \text{COD}_{max} \end{cases} \end{cases}$$

$$(2.11)$$

where γ_t is the fraction of tensile strength beyond which gel is absorbed by the crack, and Γ_r is a residual AAR retention factor for AAR under tension. If an elastic model is used, then f'_t is the tensile strength, and σ_I the maximum principal tensile stress. In contrast, if a smeared crack model is adopted, then COD_{max} would be the maximum crack opening displacement at the current Gauss point, and w_c the maximum crack opening displacement on the tensile softening curve, (Wittmann et al., 1988). Since concrete pores are seldom interconnected and the gel viscosity is relatively high, gel absorption by the pores is not explicitly taken into account. Furthermore, gel absorption by the pores is instead taken into account in the kinetic equation through the latency time, which depends on concrete porosity. The higher the porosity, the longer the latency time.

2.4.2.2 Compressive Microcracks

Retardation due to compressive stresses are accounted for in Eq. 2.1 through $\Gamma_c(\overline{\sigma}, f'_c)$. Two reasons can be cited for this retardation, the first previously addressed in Sect. 2.4.1 caused by the maximum pressure capable of being exerted by the gel. Struble and Diamond (1981) was among the earliest researchers to identify the maximum pressure (estimated at around 11 MPa). The second reason for this reduction is gel absorption by the microcracks, as induced by compressive stress. Hsu et al. (1963) showed that microcracks develop once the compressive stresses exceed $\neq 0.45 f'_c$; these microcracks may also absorb some of the gel.

$$\Gamma_c = \begin{cases} 1 & \text{if } \overline{\sigma} \leqslant 0. \quad \text{Tension} \\ 1 - \frac{e^{\beta \overline{\sigma}}}{1+(e^{\beta}-1.)\overline{\sigma}} & \text{if } \overline{\sigma} > 0. \quad \text{Compression} \end{cases} \tag{2.12}$$

$$\overline{\sigma} = \frac{\sigma_I + \sigma_{II} + \sigma_{III}}{3 f'_c} \tag{2.13}$$

Given that this expression will also reduce expansion under uniaxial or biaxial confinement, Fig. 2.10, these conditions are more directly incorporated below through the assignment of weights.

2.5 HUMIDITY

It has long been recognized that for AAR to occur, RH must be at least equal to 0.8, hence $0 < g(h) = \frac{\epsilon(t=\infty, RH=h)}{\epsilon(t=\infty, RH=1)} \leqslant 1$ is a reduction function to account for humidity. A widely accepted (albeit simplistic) model was proposed by Capra and Bournazel (1998)

$$g(h) = h^m \tag{2.14}$$

where h is the relative humidity. An alternative model was proposed by Li et al. (2004), in which a lower RH reduces the maximum expansion:

$$g(h) = ae^{b.h} \tag{2.15}$$

These two models are contrasted in Fig. 2.11(a) and shown to be nearly identical. Li et al. (2004) also altered the kinetics of the reaction through

$$\epsilon(\infty, h) = f_\infty(h_0).\epsilon(\infty, h_0) \tag{2.16}$$

$$\tau_c(h) = \alpha_c \exp(\beta_c - \gamma_c.h).\tau_c(h_0) \tag{2.17}$$

$$\tau_l(h) = \alpha_l \exp(\beta_l - \gamma_l.h).\tau_l(h_0) \tag{2.18}$$

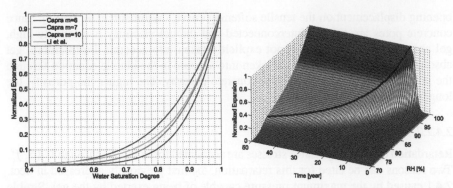

(a) Expansion only Capra and Bournazel (1998) and (b) Expansion and kinetics (at 8^oC) Li et al. (2004)
Li et al. (2004) Models

FIGURE 2.11   Effect of relative humidity

In calibrating with Larive (1998b) experiments, the coefficients are listed in Table
2.2. Fig. 2.11(b) indicates that: a) the asymptotic strain is reduced with RH, and the

TABLE 2.2

Coefficients for moisture influence on expansion and kinetics (Li et al., 2004)

a	b	α_c	β_c	γ_c	α_l	β_l	γ_l
0:0002917	8.156	0.325	9.508	8.375	0.139	10.761	8.807

asymptotic value is reached at a slower rate (as highlighted by the dark line). For
instance, at 80% RH, the asymptotic strain has been reduced by nearly 80%, and this
reduction was reached in 50 years.

Finally, in the context of a finite element analysis of a structure, s separate analyses
should be conducted in order to determine the spatial and temporal distribution of RH
inside the structure (i.e. $RH(t, x, y, z)$. While such an analysis may be warranted for
bridges of nuclear containers, it can be reasonably assumed that for dams (where RH
typically lies in the range 0f 90-95%) throughout (with the exception of a very small
layer of the exposed concrete, which can be neglected.

More refined models addressing the influence of RH on AAR expansion can be
found in Poyet et al. (2006) and Comi et al. (2012).

2.6 AAR STRAIN

In Eq. 2.1, the free *volumetric* expansion $\epsilon(\infty)$ was assumed to be volumetric, hence
the volumetric strain at time t is given by (Eq. refeq:expansion-xi)

$$\epsilon_V^{AAR}(t, x, y, z) = \xi(t, x, y, z)\epsilon(\infty) \tag{2.19}$$

Note that in general, but not always, $\epsilon(\infty)$ is uniform and does not exhibit a spatial distribution. This may not be the case however in some structures, where higher cement content was used in part of the structure (such as external concrete layers in dams) along with reactive aggregates.

Consequently, $\epsilon_V^{AAR}(t,x,y,z)$ should be "redistributed" along the three principal directions. In this model, the volumetric strain will be apportioned to the three principal stress directions[2].

The third major premise of this model is that volumetric AAR strain must be redistributed to the three principal directions according to their relative propensity to expand on the basis of a weight that is a function of the respective stresses. While the weight determination is relatively straightforward for triaxial AAR expansion under uniaxial confinement (for which some experimental data are available), it becomes less straightforward for biaxially or triaxially-confined concrete.

Mathematically speaking, the second-order engineering tensor for (small deformation) strain, at a Gauss point is defined as:

$$
E_{ij} = \underbrace{\begin{bmatrix} \epsilon_{11} & \frac{1}{2}\gamma_{12} & \frac{1}{2}\gamma_{13} \\ \frac{1}{2}\gamma_{21} & \epsilon_{22} & \frac{1}{2}\gamma_{23} \\ \frac{1}{2}\gamma_{31} & \frac{1}{2}\gamma_{32} & \epsilon_{33} \end{bmatrix}}_{\text{general}} \; ; \quad \underbrace{\begin{bmatrix} \epsilon_I & 0 & 0 \\ 0 & \epsilon_{II} & 0 \\ 0 & 0 & \epsilon_{III} \end{bmatrix} \equiv \left\{ \begin{array}{c} \epsilon_I \\ \epsilon_{II} \\ \epsilon_{III} \end{array} \right\}}_{\text{Principal Strains}}
$$

$$(2.20)$$

where we adopted the Voigt notation for the last representation of the principal strains. The vectors of anisotropic thermal strains and the yet-to-be-determined AAR strain vector are given by:

$$
\epsilon_0^{Thermal} = \left\{ \begin{array}{c} \alpha_I \Delta T \\ \alpha_{II} \Delta T \\ \alpha_{III} \Delta T \end{array} \right\} \qquad \epsilon_0^{AAR} = \left\{ \begin{array}{c} \epsilon_I^{AAR} \\ \epsilon_{II}^{AAR} \\ \epsilon_{III}^{AAR} \end{array} \right\} \qquad (2.21)
$$

Let's focus our attention on the last vector, in knowing ϵ_V^{AAR} (which is the sum of the yet-to-be-determined ϵ_I^{AAR}, ϵ_{II}^{AAR} and ϵ_{III}^{AAR}), and the principal stresses σ_I, σ_{II} and σ_{III}. We will re-label this vector of principal stresses as $\sigma_k, \sigma_l, \sigma_m$ and assign to each of the three principal directions an AAR strain proportional to the corresponding stress through a weight.

2.6.1 WEIGHTS

The use of weights will control AAR volumetric expansion distribution. For instance, with reference to Fig. 2.12, let's consider three scenarios.

---

[2]Strictly speaking, in nonlinear analysis, the principal strains directions are not co-aligned with the principal stress directions, hence this may be slightly erroneous.

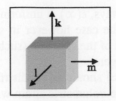

(a) Principal stresses

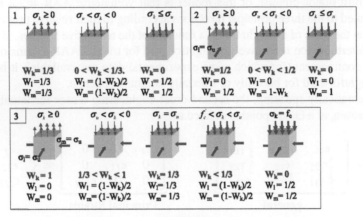

(b) Weights

FIGURE 2.12   Weight of volumetric AAR redistribution in selected cases

Uniaxial State of stress, where the three following cases are distinguished:

1. According to the first case, uniaxial tension is present; hence, the volumetric AAR strain is equally redistributed in all three directions.
2. Under a compressive stress greater than the limiting stress (σ_u), the weight in the corresponding (k) direction should be less than one-third. The remaining AAR needs to be equally redistributed in the other two directions.
3. If the compressive stress is less than σ_u, then AAR expansion in the corresponding direction is prevented (i.e. weight equal to zero), and thus the other two weights must be equal one-half.

Biaxial state of stress, in which a compressive stress equals σ_u in one of the three principal directions. In this case, the corresponding weight will always be equal to zero. As for the three possible combinations:

1. Tension in one direction, equal weights of one-half.
2. Compression greater than σ_u in one direction, then the corresponding weight must be less than one-half and the remaining weight is assigned to the third direction.
3. Compression less than σ_u, then the corresponding weight is once again zero, and a unit weight is assigned to the third direction.

Triaxial state of stress, in which σ_u acts on two of the three principal directions. The five following cases can be identified:

1. Tension along direction k, then all the expansion is along k.
2. Compressive stress greater than σ_u, yielding a triaxial state of compressive stress, and the corresponding weight will lie between one and one-third. The remaining weight complement is equally distributed in the other two directions.
3. Compression equal to σ_u, hence a perfect triaxial state of compressive stress is obtained. This case produces equal weights of one-third; it should be noted that the overall expansion is reduced through Γ_c.
4. Compression less than σ_u, but greater than the compressive strength. In this case, the weight along k should be less than one-third, with the remaining weights equally distributed along the other two directions.
5. Compression equal to the compressive strength. In this case, the corresponding weight is reduced to zero, and the other two weights are each equal to one-half.

Based on the preceding discussion, this weight allocation scheme can be generalized along direction k as follows:

1. Given σ_k, identify the quadrant encompassing σ_l and σ_m, Fig. 2.13[3]. Weight will be determined through a bilinear interpolation for these four neighboring nodes.

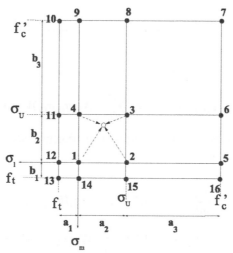

FIGURE 2.13 Weight domain

---

[3]Since compressive stresses are quite low compared to the compressive strength, we ignore the strength gained through the biaxiality or triaxiality of the stress tensor (Kupfer and Gerstle, 1973). Furthermore, the strength gain is only about 14% for equibiaxial compressive stresses, (CEB, 1983).

2. Determine the weights of the neighboring nodes from Table 2.3 through a proper linear interpolation of σ_k.

TABLE 2.3
Triaxial Weights

No.	Node σ_l	Node σ_m	$\sigma_k \geqslant 0$	Weights $\sigma_k = \sigma_u$	Weights $\sigma_k = f'_c$
1	0.	0.	1/3	0.	0.
2	σ_u	0.	1/2	0.	0.
3	σ_u	σ_u	1.	1/3	0.
4	0.	σ_u	1/2	0.	0.
5	f'_c	0.	1/2	0.	0.
6	f'_c	σ_u	1.	1/2	0.
7	f'_c	f'_c	1.	1.	1/3
8	σ_u	f'_c	1.	1/2	0.
9	0.	f'_c	1/2	0.	0.
10	f'_t	f'_c	1/2	0.	0.
11	f'_t	σ_u	1/2	0.	0.
12	f'_t	0.	1/3	0.	0.
13	f'_t	f'_t	1/3	0.	0.
14	0.	f'_t	1/3	0.	0.
15	σ_u	f'_t	1/2	0.	0.
16	f'_c	f'_t	1/2	0.	0.

3. Compute the weight from:

$$W_k(\sigma_k, \sigma_l, \sigma_m) = \sum_{i=1}^{4} N_i(\sigma_l, \sigma_m) W_i(\sigma_k) \qquad (2.22)$$

where N_i is the typical two bilinear shape function used in finite element computations, given by:

$$\mathbf{N}(\sigma_l, \sigma_m) = \frac{1}{ab} \lfloor (a - \sigma_l)(b - \sigma_m) \quad \sigma_l(b - \sigma_m) \quad \sigma_l \sigma_m \quad (a - \sigma_l)\sigma_m \rfloor \qquad (2.23)$$

$$\mathbf{W}(k) = \lfloor W_1(\sigma_k) \quad W_2(\sigma_k) \quad W_3(\sigma_k) \quad W_4(\sigma_k) \rfloor^t \qquad (2.24)$$

$$a = (a_1|a_2|a_3) \qquad b = (b_1|b_2|b_3) \qquad (2.25)$$

$$\sigma_l = (\sigma_l|f'_c - \sigma_l) \qquad \sigma_m = (\sigma_m|f'_c - \sigma_m) \qquad (2.26)$$

The $i - j$ stress space is decomposed into nine distinct regions, Fig. 2.13, where σ_u is the upper (signed) compressive stress below which no AAR expansion can occur along the corresponding direction (except in triaxially-loaded cases). Hence, a and b are the dimensions of the quadrant inside which σ_i and σ_j reside.

Weights of the individual nodes are, in turn, interpolated according to the principal stress component in the third direction σ_k, Table 2.3. It should be noted that these weights are, for the most part, based on the work of Larive (1998b) and Multon (2004), but in some cases due to a lack of sufficient experimental data they are based on simple "engineering common sense". A straightforward example of the weight evaluation is shown below.

Assuming that the principal stresses are given by $\lfloor \sigma_l \quad \sigma_m \quad \sigma_k \rfloor = \lfloor -5.0 \quad -8.0 \quad -5.0 \rfloor$ MPa, and that f_c, f_t' and σ_u are equal to -30.0, 2.0, and -10.0 MPa, respectively, we seek to determine W_k.

The stress tensors place us inside the quadrant defined by nodes 1-2-3-4, whose respective weights are equal to: $W_1 = \frac{1}{2}\left(\frac{1}{3}\right) = \frac{1}{6}$, $W_2 = \frac{1}{2}\left(\frac{1}{2}\right) = \frac{1}{4}$, $W_3 = \frac{1}{3} + \frac{1}{2}\left(1.0 - \frac{1}{3}\right) = \frac{2}{3}$, and $W_4 = \frac{1}{2}\left(\frac{1}{2}\right) = \frac{1}{4}$ a and b are both equal to -10 MPa, and the "shape factors" will be: $N_1 = \frac{1}{100}\left[(-10 + 5)(-10 + 8)\right] = \frac{1}{10}$, $N_2 = \frac{1}{100}\left[-5(-10 + 8)\right] = \frac{1}{10}$, $N_3 = \frac{1}{100}\left[(-5)(-8)\right] = \frac{4}{10}$, $N_4 = \frac{1}{100}\left[-8(-10 + 5)\right] = \frac{4}{10}$, and finally $W_k = \frac{1}{10} \times \frac{1}{6} + \frac{1}{10} \times \frac{1}{4} + \frac{4}{10} \times \frac{2}{3} + \frac{4}{10} \times \frac{1}{4} = 0.40833$

Based on the earlier work of Struble and Diamond (1981), where it was reported that no gel expansion can occur at pressures above 11 MPa (though for a synthetic gel), σ_u is set as -10 MPa. This value was also confirmed by Larive (1998b). f_t' and f_c' are the concrete tensile and compressive strengths, respectively.

2.6.2 AAR LINEAR STRAINS

The volumetric strain at a Gauss point and at time t will be given by (Eq. 2.19) $\epsilon_V^{AAR}(t, x, y, z) = \xi(t, x, y, z)\epsilon(\infty)$, and individual strains can now be obtained from

$$\epsilon_i^{AAR}(t, x, y, z) = W_i \dot{\xi}(t, x, y, z)\epsilon(\infty) \tag{2.27}$$

and the resulting relative weights are shown in Fig. 2.14.

It should be noted that the proposed model will indeed result in an anisotropic AAR expansion. While not explicitly expressed in tensor form, the anisotropy stems from the various weights assigned to each of the three principal directions.

2.6.3 DETERIORATION

Since this deterioration is time-dependent, the following time-dependent nonlinear model is considered, Fig. 2.15.

$$E(t, T) = E_0\left[1 - (1 - \beta_E)\,\xi(t, T)\right] \tag{2.28}$$
$$f_t'(t, T) = f_{t,0}'\left[1 - (1 - \beta_f)\,\xi(t, T)\right] \tag{2.29}$$

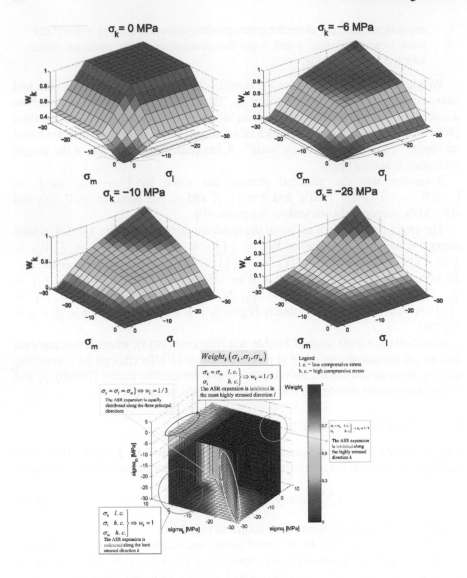

FIGURE 2.14 Relative Weights

where E_0 and $f'_{t,0}$ are the original elastic modulus and tensile strength, β_E and β_f are the corresponding residual fractional values when ϵ_{AAR} tends to $\epsilon_{AAR}(\infty)$.

Finally, the possible decrease in compressive strength with AAR has been ignored. Most of the literature focusing on the mechanical properties of concrete subjected to AAR show little evidence of a decrease in compressive strength (as would be expected since the stresses will essentially be closing the AAR-induced cracks). Furthermore,

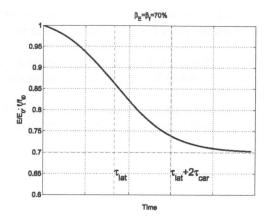

FIGURE 2.15   Degradation of E and f_t'

in dams (of both the gravity and arch types), compressive stresses lie well below the compressive strength, which is quite different from tensile stresses.

2.7 SUMMARY

This chapter has presented in great detail the AAR constitutive model proposed by the author. Aside from the fact that it addresses all major (and modern) requirements for AAR modeling, its greatest advantage is its relatively simple implementation within existing linear or nonlinear finite element codes (examples of two such implementations in third party software can be found in Sect. 7.3 and 7.4). For the analysis of complex 3D structures, it may indeed be preferable to perform a linear elastic analysis, which also properly accounts for AAR. This option may not be feasible with most existing codes, since modeling AAR forces the user to adopt a nonlinear model, which may (at least at first) severely complicate the analysis.

The major points to be retained from this chapter include:

1. AAR requires a relative humidity of 80% to be initiated, as the reaction rate increases with temperature.
2. The net result of this reaction is the formation of localized cracks, which are filled with the AAR gel.
3. The kinetics of this reaction follow an "S"-shaped curve, i.e. it starts out very slow, then increases rapidly and finally slows down gradually. As opposed to the maximum AAR-induced strain, the reaction rate is temperature-dependent.
4. The manifestation of this reaction is inhibited by compressive stresses greater than approx. 10 MPa, and expansion is redirected in the orthogonal directions. At about 20 MPa, expansion in the stress direction has practically ceased (due to the microcracks).

5. It is indicated that under a triaxial state of stress (irrespective of the volumetric stress), the reaction continues with a constant, albeit reduced, volumetric strain.
6. Due to the stress state, this reaction will result in an anisotropic expansion.
7. A reduction occurs in both tensile strength and elastic modulus associated with the microcracks caused by AAR.
8. The proposed model can be easily implemented in existing finite element codes as an additional initial strain ϵ_0 very analogous to temperature strain.

3 Constitutive Model; Concrete

Chapter written with Dr. Jan Cervenka

As highlighted in Section 1.5.2, most existing model designs tightly couple the concrete AAR model with its constitutive counterpart (which relates incremental strain to incremental stress). The only major exception is this author's model, according to which the AAR expansion model is completely independent of the concrete constitutive model. This approach features two major advantages: relative ease of model implementation in existing codes; and a streamlined approach to conducting a simple linear (elastic) analysis (offering the advantage of speed), albeit with a rather "sophisticated" AAR model.

Having already described the AAR model in Chapter 2, this chapter will first address the critical issue of concrete craking. It will then proceed into discussing two finite element approaches to model (primarily) cracking, (more specifically D_t in Eq. 1.20): a continuum based one rooted in the theory of plasticity, and a second one rooted in fracture mechanics. The former is implemented in a so-called "smeared crack model", whereas the later is implemented in a so-called "discrete crack model".

Important note: A reading and understanding of this chapter is not essential for implementing an initial AAR numerical model. The model previously described is in fact capable of performing reasonably well, with nothing more sophisticated than a continuous linear elastic model. This chapter would only be relevant for analysts seeking to couple the AAR model with a nonlinear concrete model.

3.1 INTRODUCTION

Since dams are the structures that most visibly suffer from AAR, it is worth noting that historically dams have served as catalysts for structural analysis innovations. While it is widely known that the first reported application of the finite element method in civil engineering was indeed the fracture analysis of a (cracked) concrete dam by Sims et al. (1964), few are aware that one of the first finite differences applications was also on masonry dams (Richardson, 1911), not to mention the numerous examples of dams in the first edition of Zienkiewicz (1967).

At the outset, concrete cracks were naturally (and intuitively) modeled as what would come to be known as *discrete* cracks, whereby a mesh discontinuity accompanied the crack. In other words, if cracking were the sole concern, then the element **D** would be constant but the mesh would require continuous adjustments. This was the emphasis in the early work by Ngo and Scordelis (1967) and Nilson (1968). A major disadvantage of this approach however was the need to continuously "re-mesh" in order to accommodate crack propagation. One solution to this problem was first proposed by Rashid (1968), who introduced the concept of a *smeared* crack (for the analysis of nuclear reactors). According to this approach, the mesh is stationary, while cracking is taken into account by continuously modifying D_t (the tangential material matrix). Since this breakthrough effort, many finite element models have been developed for concrete based on this smeared crack approach. Yet a number of major challenges remain, namely: a) how to prevent mesh size/orientation dependency; b) how to ensure localization (as opposed to a "cloud" of smeared cracks that only vaguely resemble a major physical crack); and c) how to avoid the use of unrealistically small elements.

During the early stages, the author's Ph.D. thesis (Saouma, 1980) revisited the discrete crack model in automating the re-meshing procedure, (Ingraffea, 1977), which indeed proved to be a viable approach for those cases where a few major structural cracks (or joints) dictated the response of a structure (as is the case with unreinforced concrete). This model has been successfully applied on many occasions (Saouma, 2013a).

It should be mentioned that the nonlinear modeling of reinforced concrete is not yet considered a solved problem. The smeared crack approach still suffers from the dual difficulty of "localizing" a crack and avoiding bias due to mesh orientation. Ironically, this approach has revisited the discrete crack model in an effort to incorporate its benefits, (de Borst, R. and Remmers, J.C. and Needleman, A. and Abellan, M.A., 1989); moreover, based on various "Benchmarks", or (blind) "Round-Robins", the analysis of even a simple reinforced concrete beam can yield results with an acceptable 10% normalized standard deviation among what are perceived to be "good" models.

This chapter will present two models, one based on the smeared crack approach (or a nonlinear continuum model) the other based on the discrete crack model. The first model (Cervenka and Papanikolaou, 2008) will be used for reinforced concrete structures, where the multitude of cracks precludes the application of discrete crack modeling, for which case compressive stresses may exceed about $0.45 f_c'$ and reinforcements may yield. The second (Cervenka et al., 1998) model will be introduced when just a few major structural cracks (or joints) are present, such as in dams, see Fig. 3.1. Both models have been successfully implemented in the context of AAR analysis by the author and have served as examples of constitutive models for eventual coupling with the previously presented AAR model. In addition, both models have been successfully implemented in the finite element code Saouma et al. (2010).

3.2 NONLINEAR RESPONSE OF CONCRETE

Before proceeding with the development of continuum and fracture-based models, it is important to review once again the concrete response under tension, compression and shear, though this time from a nonlinear perspective.

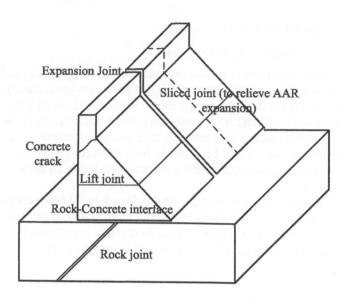

FIGURE 3.1 Example of a crack/joints/faults capable of being modeled by a discrete crack

3.2.1 CONCRETE IN TENSION

Let's consider a concrete specimen loaded uniaxially by a tensile load F; its elongation can be recorded by an LVDT (linearly variable displacement transducer) mounted in configuration 1, 2 or 3, see Figs. 3.2 and 6.2.

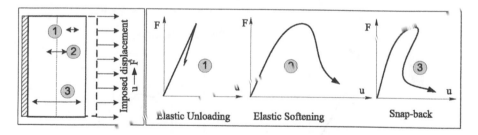

FIGURE 3.2 Capturing localization in a concrete plate subjected to increased imposed displacement

Once the peak load has been reached and cracking is initiated, transducer 1 will record the elastic unloading, transducer 2 the strain softening, and 3 the snap-back. Transducer 2 measurements, which include the crack surface area, indicate that the fracturing process is not a sudden event, but rather a gradual release of stress along the crack surfaces. This process corresponds to the initiation of micro-cracks, their gradual propagation and lastly the formation of the final macro-crack. This process constitutes the essence of Hillerborg's model, which will be described below.

3.2.2 HILLERBORG'S MODEL

3.2.2.1 σ-COD Diagram, Hillerborg's Model

From the previous discussion, it is clear that concrete softening can be characterized by a stress-crack opening width curve (as opposed to a stress-strain curve). An exact characterization of the softening response should ideally be obtained from a uniaxial test of an uncracked specimen. It has been determined however Li and Liang (1986); Hordijk et al. (1989) that not only are such tests extremely sensitive, but radically different test results may indeed be obtained from different geometries, sizes and testing devices. Consequently, the softening curve is often indirectly derived by testing notched specimens.

In what is probably the most widely referenced work in the literature on the non-linear fracture of concrete, Hillerborg Hillerborg et al. (1976) presented in 1976 a very simple and elegant model, which has been described qualitatively in previous publications. In this model, the crack is composed of two parts (Fig. 3.3):

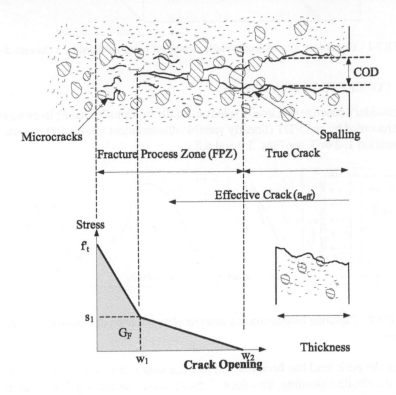

FIGURE 3.3   Hillerborg's fictitious crack model

1. A real, or physical, crack across which no stresses can be transmitted. This zone contains both displacement and stress discontinuities.

2. A fictitious crack, or a Fracture Process Zone (FPZ), ahead of the previous zone, as characterized by:

    a. peak stress at its tip equal to the tensile strength of concrete
    b. decreasing stress distribution from f_t' at the tip of the fictitious crack to zero at the tip of the physical crack

    It should be noted that both displacement and stress continuities exist along the FPZ.

This model is among the most widely used in the finite element analysis of non-linear fracture mechanics; however, due to its computational complexity, few "engineering" structures have actually been analyzed. In addition:

1. An inflection point is located in the descending branch.

    a. The first part has been associated with (disconnected) microcracking ahead of the stress-free crack
    b. The second part has been associated with crack bridging by aggregates.
2. The area under the curve, referred to as fracture energy G_F (not to be confused with G_c or the critical energy release rate), provides a measure of the energy needed to be spent in order to generate a unit surface area of crack.
3. Shape of the softening diagram ($\sigma - COD$), with in general a bilinear model capable of being used for the strain softening. One topic of considerable research lately is the experimental determination of fracture energy G_F and the resulting shape of the softening diagram (Cedolin et al., 1987; Petersson, 1981; Wittmann et al., 1988; Jeang and Hawkins, 1985; Gopalaratnam and Shah, 1985; Duda, 1990; Giuriani and Rosati, 1986). To assess the relevance of knowing the precise value of G_F and the softening curve shape in numerical simulations, three different sets of fracture experiments will be analyzed using the average reported fracture energy. The shape of the softening diagram is assumed to be the bilinear one proposed in Wittmann et al. (1988), Fig. 3.3. This simple model can be uniquely defined in terms of tensile strength f_t' and fracture energy G_F. In Brühwiler and Wittmann (1990), it was found that the optimal points for concrete with a 1" maximum size aggregate are:

$$s_1 = 0.4 f_t' \tag{3.1}$$

$$w_1 = 0.8 \frac{G_F}{f_t'} \tag{3.2}$$

$$w_2 = 3 \frac{G_F}{f_t'} \tag{3.3}$$

whereas for structural concrete, (Wittmann et al., 1988), the corresponding
values are:

$$s_1 = \frac{f'_t}{4} \tag{3.4}$$

$$w_1 = 0.75\frac{G_F}{f'_t} \tag{3.5}$$

$$w_2 = 5\frac{G_F}{f'_t} \tag{3.6}$$

where f'_t is the uniaxial tensile strength. In the context of a nonlinear fracture mechanics analysis, this tensile strength cannot be set to zero, which
would effectively eliminate the fracture process zone. Since f'_t is seldom
determined experimentally, it is assumed herein to equal 9% of f'_c, Mindess
and Young (1981).

4. Instead of a direct tensile test, a flexural test can be performed under strain
   control, allowing the fracture energy G_F to still be determined from the area
   under the load and corresponding displacement curve.

5. For dynamic analyses, the fracture properties of dam concrete depend on the
   rates of both loading and preloading. Test results Brühwiler and Wittmann
   (1990) show that these fracture properties generally increase with the loading
   rate; however, dynamic compressive preloading leads to a reduction in the
   fracture properties at quasi-static and high loading rates.

A number of empirical equations for concrete fracture energy G^F_I have been proposed. The following one was suggested by METI (2001):

$$(0.79d_a + 80)\left(\frac{f_c}{10}\right)^{0.7} \tag{3.7}$$

where d_a is the maximum aggregate size in mm, and f_c the compressive strength in
MPa.

An earlier model by Bažant, Z.P. (1984) found that G_F may be predicted (with a
coefficient of variation of roughly 16%) from the following empirical equation:

$$G_F = 0.0214(f'_t + 127)f'^2_t\frac{d_a}{E_c} \tag{3.8}$$

where E_c and f'_t are expressed in pounds per square inch, and d_a is the aggregate size
in inches. In contrast, through the use of extensive nonlinear optimization techniques
based on the Levenberg-Marquardt algorithm, Bažant and Becq-Giraudon (2001)
obtained two simple approximate formulae for the means of G_f and G_F as functions
of compressive strength f'_c, maximum aggregate size d_a, water-cement ratio w/c and

aggregate shape (i.e. crushed or river):

$$G_f = \alpha_0 \left(\frac{f'_c}{0.051}\right)^{0.46} \left(1 + \frac{d_a}{11.27}\right)^{0.22} \left(\frac{w}{c}\right)^{-0.30} \qquad \omega_{G_f} = 17.8\%$$

$$G_F = 2.5\alpha_0 \left(\frac{f'_c}{0.051}\right)^{0.46} \left(1 + \frac{d_a}{11.27}\right)^{0.22} \left(\frac{w}{c}\right)^{-0.30} \qquad \omega_{G_F} = 29.9\%$$

$$c_f = \exp\left[\gamma_0 \left(\frac{f'_c}{0.022}\right)^{-0.019} \left(1 + \frac{d_a}{15.05}\right)^{0.72} \left(\frac{w}{c}\right)^{0.2}\right] \qquad \omega_{c_f} = 47.6\%$$

$$(3.9)$$

According to this set-up, $\alpha_0 = \gamma_0 = 1$ for rounded aggregates, while $\alpha_0 = 1.44$ and $\gamma_0 = 1.12$ for crushed or angular aggregates; ω_{G_f} and ω_{G_F} are the coefficients of variation of the ratios G_f^{test}/G_f and G_F^{test}/G_F, for which a normal distribution may be assumed; and ω_{c_f} is the coefficient of variation of c_f^{test}/c_f, for which a lognormal distribution is advised (ω_{c_f} is approximately equal to the standard deviation of $\ln c_f$).

For gravity dams, the author has chosen a value of 1.35×10^{-3} kip/in, (Saouma et al., 1991b). Note that for arch dams, this value could probably be increased, on the basis of laboratory test results (Brühwiler, E., 1988).

3.2.2.2 Localization

In light of the above discussion, we have reexamined the plate in Figure 3.2 subjected to an imposed displacement. The same plate is shown once again in Figure 3.4(a), while localization of the strain due to increasingly imposed displacements is displayed in Figure 3.4(b). A uniform strain field is initially apparent, yet at some point (i.e. an imposed displacement of 70 units) localization starts and then becomes sharply defined for an imposed displacement of 100 units. The gradual increase in strain, localization at the formed crack and the elastic return outside the crack are best illustrated in Figure 3.4(c).

3.2.3 CONCRETE IN COMPRESSION

The compressive behavior of concrete is characterized by a rather high compressive strength compared to its tensile strength, as previously discussed in Section 3.2.1. For typical concrete specimens, the compressive strength is about 10 times higher than tensile strength. The concrete in compression behaves at first like a linear material, but relatively quickly thereafter its response departs from the linear curve: this is due to the development of micro-cracks both in the cement matrix and around concrete aggregates. As the applied stress approximates the compressive strength, a highly nonlinear response is usually captured. Once the compressive strength has been reached, i.e. in the post-peak gradual softening, a response is obtained similar to that observed in tension (see Section 3.2.1). The typical response of a uniaxial compression experiment is shown in Figure 3.5; it has been experimentally observed by van Mier (1984) that the displacement w_d during the softening slope becomes independent of specimen dimension.

In a multiaxial state of stress, the compressive strength may increase even if a sufficient confining stress is applied in the other directions, as depicted in Figure

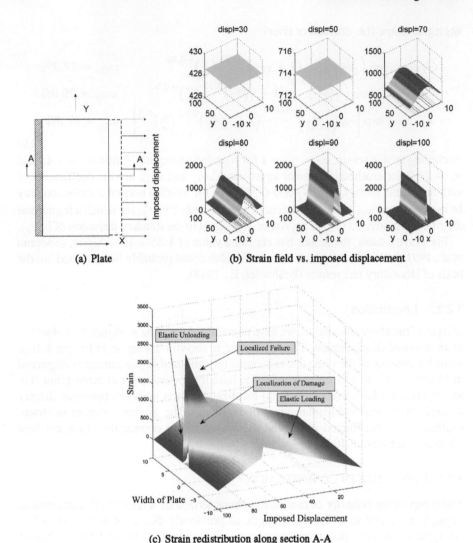

(a) Plate

(b) Strain field vs. imposed displacement

(c) Strain redistribution along section A-A

FIGURE 3.4   Localization of cracks in concrete and accompanying strain redistribution

3.6a. The failure surface in a three-dimensional space of principal stresses for a typical concrete thus resembles a cone with a nearly triangular cross-section, as described in 3.6b.

3.2.4   CONCRETE IN SHEAR

Considerable experimental effort has been devoted to investigating concrete properties in shear. Van Mier et al. (1992) defined shear failure as the narrow zone of inclined tensile cracks remaining in the plane of applied shear loads.

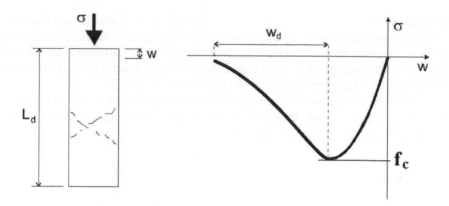

FIGURE 3.5   Uniaxial compression experiment

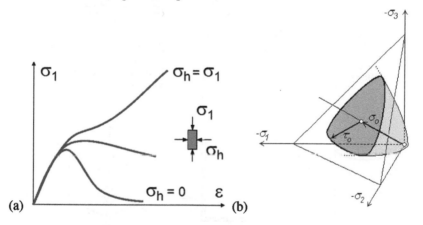

FIGURE 3.6   Concrete behavior under a multiaxial state of stress: a) Uniaxial test with confinement, and b) Typical 3D failure criterion for concrete

Various shear test geometries are shown in Figure 3.7. One of the first experiments was conducted by Fenwick and Paulay (1968). During these early experiments, a cracked specimen was subjected to a shear force while the crack opening was held constant. Shear was introduced once the specimen had been fully cracked; consequently, these experiments provide information on the shear properties of cracked concrete. The ultimate failure can nevertheless be attributed to flexural cracking rather than sliding along the pre-formed crack. An experimental set-up similar to that used by Paulay and Loeber (1987) was adopted by Walraven (1980) in order to derive empirical relationships between stresses and relative crack displacements.

Other experimental set-ups have been proposed by Bazant and Pfeifer (1986), Swartz et al. (1988), Swartz and Taha (1990), Ballatore et al. (1990), Van Mier et al. (1992), Keuser and Walraven (1989), B. (1992) and Hassanzadeh (1992) to determine the shear fracture properties of concrete. As pointed out by Ingraffea and Panthaki (1985), in these tests crack propagation is mainly due to tensile cracking, as opposed to shear. In reality, pure shear failure in plain concrete is very difficult to obtain and can only occur under the following conditions: 1) triaxial compression, 2) dynamic loading, and 3) fiber-reinforced concrete (Van Mier et al. (1992)).

It should be pointed out however that shear properties, i.e. strength and stiffness, along a cracked surface may be significant for some problems, such as cyclic loading or concrete structures with reinforcement.

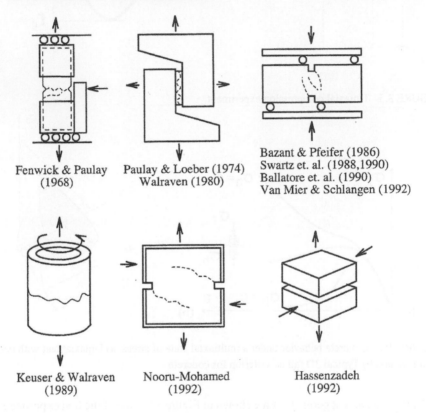

Fenwick & Paulay (1968)

Paulay & Loeber (1974) Walraven (1980)

Bazant & Pfeifer (1986)
Swartz et. al. (1988,1990)
Ballatore et. al. (1990)
Van Mier & Schlangen (1992)

Keuser & Walraven (1989)

Nooru-Mohamed (1992)

Hassenzadeh (1992)

FIGURE 3.7   Various test geometries for shear fracture

3.3   THE NONLINEAR CONTINUUM MODEL

This first version, developed by Cervenka and Papanikolaou (2008), is a combined fracture-plastic model built from constitutive models for both tensile (fracturing) and compressive (plastic) behavior. Such a fracture model is based on the classical orthotropic smeared crack formulation and crack band model; it employs a Rankine failure criterion, includes exponential (or user-defined) softening and moreover can serve as either a rotated or fixed crack model. The hardening/softening plasticity model is based on the Menétrey and Willam (1995) failure surface. Both models use a return mapping algorithm for integrating the constitutive equations. Special attention is paid to developing an algorithm that combines the two models. This combination algorithm is based on a recursive substitution technique and allows each model to be developed and formulated separately. The algorithm can handle cases with active failure surfaces for both models, as well as in the presence of physical changes such as crack closure. Moreover, the model may be used to simulate concrete cracking, crushing under high confinement and crack closure due to crushing in other material directions.

The method of strain decomposition, as introduced by de Borst (1986), is used to associate fracture and plasticity models, both of which have been developed within the framework of the return mapping algorithm by Wilkins (1964). This approach guarantees the solution for all strain increment magnitudes. From an algorithmic standpoint, this problem is then transformed into finding an optimal return point on the failure surface. The combined algorithm must determine the separation of strains into plastic and fracturing components while preserving the stress equivalence in both models. The proposed algorithm is based on a recursive iterative scheme. It has been proven that such a recursive algorithm cannot reach convergence in certain cases, e.g. softening and dilating materials. For this reason, the recursive algorithm has been extended by a variation of the relaxation method in order to stabilize convergence.

3.3.1 MATERIAL MODEL FORMULATION

The material model formulation assumes small strains along with their decomposition into elastic ε_{ij}^e, plastic ε_{ij}^p and fracturing components ε_{ij}^f, (de Borst, 1986).

$$\dot{\varepsilon}_{ij} = \dot{\varepsilon}_{ij}^e + \dot{\varepsilon}_{ij}^p + \dot{\varepsilon}_{ij}^f \qquad (3.10)$$

In nonlinear analysis, the solution proceeds step by step, whereby at each step a new strain vector increment is calculated and then used to evaluate the new stress state, i.e.:

$$\sigma_{ij}^n = \sigma_{ij}^{n-1} + E_{ijkl}(\dot{\varepsilon}_{kl} - \dot{\varepsilon}_{kl}^p - \dot{\varepsilon}_{kl}^f) \qquad (3.11)$$

where the plastic strain rate of change $\dot{\varepsilon}_{kl}^p$ and fracturing strain rate $\dot{\varepsilon}_{kl}^f$ must be evaluated in the plastic and fracturing model, respectively. The flow rule governing the evolution of plastic and fracturing strain can be summarized by the following two equations:

$$\text{Plastic model}: \quad \dot{\varepsilon}_{ij}^p = \dot{\lambda}^p \cdot m_{ij}^p, \quad m_{ij}^p = \frac{\partial g^p}{\partial \sigma_{ij}} \qquad (3.12)$$

$$\text{Fracture model}: \quad \dot{\varepsilon}_{ij}^f = \dot{\lambda}^f \cdot m_{ij}^f, \quad m_{ij}^f = \frac{\partial g^f}{\partial \sigma_{ij}} \qquad (3.13)$$

where $\dot{\lambda}^p$ is the plastic multiplier rate and g^p the plastic potential function. It is possible to define analogous quantities (see Carol (1994)) for the fracturing model, i.e. $\dot{\lambda}^f$ is the inelastic fracturing multiplier and g^f the potential defining the direction of inelastic fracturing strains within the fracturing model. Consistency conditions can then be used to evaluate the change in both plastic and fracturing multipliers.

$$\dot{f}^p = n_{ij}^p \cdot \dot{\sigma}_{ij} + H^p \cdot \dot{\lambda}^p = 0, \quad n_{ij}^p = \frac{\partial f^p}{\partial \sigma_{ij}} \tag{3.14}$$

$$\dot{f}^f = n_{ij}^f \cdot \dot{\sigma}_{ij} + H^f \cdot \dot{\lambda}^f = 0, \quad n_{ij}^f = \frac{\partial f^f}{\partial \sigma_{ij}} \tag{3.15}$$

This approach yields a system of two equations for the two unknown multiplier rates $\dot{\lambda}^p$ and $\dot{\lambda}^f$ and is analogous to the multi-surface plasticity problem Simo et al. (1988). The combination of these two models will be described in detail in Section 3.3.4.

3.3.2 RANKINE-FRACTURING MODEL FOR CONCRETE CRACKING

The tensile behavior in concrete can be very well described by linear behavior up to the point where one of the principal stresses reaches the tensile strength value. This point also initiates the development of micro-cracks, which gradually propagate and coalesce into a macro-crack. The crack initiation process can be accurately described by the Rankine criterion for each principal stress direction ($k = 1, 2, 3$).

$$f_k^f = {}^t\sigma_{ij} \cdot n_i^k \cdot n_j^k - f_t \leqslant 0 \tag{3.16}$$

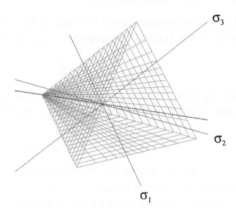

FIGURE 3.8   Rankine Criterion represented in a 3D stress space of principal stresses

The Rankine criterion 3.16 actually comprises three distinct planes forming a pyramid in the stress space, as depicted in Figure 3.8. It is assumed herein that strains and stresses are converted into material directions given by the vectors $\mathbf{n}^k$, which correspond to the crack directions. According to this model, a maximum of three independent crack directions are possible, with these directions being perpendicular to one another. It is possible to identify two models for the definition of crack directions n^k. Each direction may be "fixed" to the principal stress direction where the stress exceeds tensile strength f_t; as a result, this model is often called the "fixed crack model". An alternative model would be the "rotated crack model", in which the crack directions are always aligned with the directions of maximum principal strains. Therefore, is identified with the trial stress and with the concrete tensile strength.

The trial stress ${}^t\sigma_{ij}$ state for the fracture model is computed by the elastic predictor as follows:

$$ {}^t\sigma_{ij} = {}^n\sigma_{ij} + D_{ijkl} \cdot d\varepsilon_{kl} \tag{3.17} $$

If the trial stress ${}^t\sigma_{ij}$ does not satisfy the Rankine failure criterion 3.16, then the fracture strain increment in direction k can be calculated under the assumption that the final stress state must satisfy the following equation:

$$ f_k^f = {}^{n+1}\sigma_{ij} \cdot n_i^k \cdot n_j^k - f_t = {}^t\sigma_{ij} \cdot n_i^k \cdot n_j^k - D_{ijmn} \cdot d\varepsilon_{mn}^f \cdot n_i^k \cdot n_j^k - f_t = 0 \tag{3.18} $$

This equation can be further simplified under the assumption that the fracturing strain increment is normal to the Rankine failure surface and moreover that never more than one direction is being checked each time. For failure surface k, the fracturing strain increment has the following form (associated form, i.e. $g^f = f^f$):

$$ d\varepsilon'^{fk}_{ij} = d\lambda^k \cdot \frac{\partial f_k^f}{\partial \sigma_{ij}} = d\lambda^k \cdot n_i^k \cdot n_j^k \tag{3.19} $$

After substituting Equation 3.19 into Equation 3.18, a formula for the fracturing multiplier increment is derived:

$$ d\lambda^k = \frac{{}^t\sigma_{ij} \cdot n_i^k \cdot n_j^k - f_t\left(w^k\right)}{D_{ijmn} \cdot n_i^k \cdot n_j^k \cdot n_m^k \cdot n_n^k} \tag{3.20} $$

where:

$$ w^k = L_t \cdot \left(\hat{\varepsilon}_k^f + d\lambda^k\right) \tag{3.21} $$

The system of equations 3.20 and 3.21 must be solved iteratively since for softening materials the value of the current tensile strength $f_t(w^k)$ is a function of the crack opening (w^k), which in turn is based on the following empirical formula suggested by Hordijk (1991):

$$ \frac{\sigma}{f_t} = \left[1 + \left(c_1 \cdot \frac{w}{w_o}\right)^3\right] \cdot e^{-c_2 \cdot \frac{w}{w_o}} - \frac{w}{w_o} \cdot \left(1 + c_1^3\right) \cdot e^{-c_2} \tag{3.22} $$

where: σ is the tensile concrete stress normal to the crack, f_t the concrete tensile strength, $c1 = 3$, $c2 = 6.93$ and $w_o = 5.14 \cdot \frac{G_f}{f_t}$ (G_f is the fracture energy of the material). L_t is the characteristic dimension of the element, as introduced by Bažant and Oh (1983). The crack opening w_k is computed from the total cumulative value of fracturing strain $\hat{\varepsilon}_k^f$ in direction k, plus the current increment of fracturing strain $d\lambda$. Next, this sum is multiplied by the characteristic length L_t, which as a crack band size was introduced by Bažant, Z.P. and Oh, B.H. (1983). Various methods have been proposed for the crack band size calculation within the finite element method framework. Some authors Feenstra (1993) suggest an approach based on integration point volume, which is not well suited for distorted elements. A consistent and rather complex approach was proposed by Olivier (1989); according to the presented model, the crack band size is calculated as a width or size of the element projected onto direction k 3.9. This approach proves satisfactory for low-order, linear finite elements and will provide conservative results even for higher-order elements. In addition, the approach performs well for elements with aspect ratios different from 1, as well as for cases where the crack direction is not aligned with the finite element edges. It can be

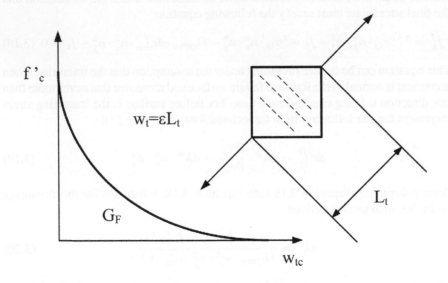

FIGURE 3.9   Tensile softening and characteristic length, (Cervenka and Papanikolaou, 2008)

demonstrated that by expanding $f'^t(w_k^{max})$ into a Taylor series, this iteration scheme converges as long as:

$$\left| -\frac{\partial f'^t(w_k^{max})}{\partial w} \right| < \frac{E_{kkkk}}{L_t} \tag{3.23}$$

A distinction is to be drawn between the total maximum fracturing strain during loading $\hat{\varepsilon}_{kk}^{'f}$ and the current fracturing strain $\varepsilon_{ij}^{'f}$, which is determined according to Rots and Blaauwendraad (1989) in matrix form:

$$\varepsilon'^f = (\mathbf{D} + \mathbf{D'}^{cr})^{-1} \cdot \mathbf{D} \cdot \varepsilon \quad \sigma'_{ij} = D'^{cr}_{ijkl} \cdot \varepsilon'^f_{kl} \tag{3.24}$$

where D'^{cr}_{ijkl} is the cracking stiffness in the local material (i.e. prime) direction. No interaction is assumed between the normal and shear components; the crack tensor is thus given by:

$$Dr_{ijkl}^{cr} = 0 \text{ for } i \neq k \text{ and } j \neq l \tag{3.25}$$

The mode I crack stiffness is expressed as:

$$Dr_{iiii}^{cr} = \frac{f'^{t}(w_i^{max})}{\hat{\varepsilon}_{ii}^{'f}} \text{ (no index summation)} \tag{3.26}$$

and mode II and III crack stiffness values are assumed to equal:

$$Dr_{ijij}^{cr} = \frac{r_g^{ij} G}{1 - r_g^{ij}} \text{ (no index summation)} \tag{3.27}$$

where $i \neq j$, $r_g^{ij} = \min(r_g^i, r_g^j)$ are the minimum shear retention factors on cracks for directions i and j, as given by Kolmar (1986):

$$r_g^i = \frac{-\ln\left(\frac{\varepsilon_{ii}'}{c_1}\right)}{c_2} \tag{3.28}$$

$$c_1 = 7 + 333(\rho - 0.005) \tag{3.29}$$

$$c_2 = 10 - 167(\rho - 0.005) \tag{3.30}$$

where ρ is the reinforcement ratio, with the assumption that its value lies below 0.002. G is the elastic shear modulus. For special cases prior to the onset of cracking, i.e. as the expressions approach infinity, large penalty numbers are introduced for crack stiffness values. The shear retention factor is used only in the case of the fixed crack option.

Furthermore, the secant constitutive matrix in the material direction is analogous to one presented by Rots and Blaauwendraad (1989)

$$\mathbf{E}'^{s} = \mathbf{E} - \mathbf{E}(\mathbf{E}'^{cr} + \mathbf{E})^{-1}\mathbf{E} \tag{3.31}$$

which should then be transformed into the global coordinate system $\mathbf{E}^s = \Gamma_\varepsilon^T \mathbf{E}'^s \Gamma_\varepsilon$ where Γ_ε is the strain vector transformation matrix (i.e. global-to-local strain transformation matrix).

3.3.3 PLASTICITY MODEL FOR CONCRETE CRUSHING

Starting from the predictor-corrector formula, the stress is determined from:

$$\sigma_{ij}^n = \sigma_{ij}^{n-1} + E_{ijkl}(\Delta\varepsilon_{kl} - \Delta\varepsilon_{kl}^p) = \sigma_{ij}^t - E_{ijkl}\Delta\varepsilon_{kl}^p = \sigma_{ij}^t - \sigma_{ij}^p \tag{3.32}$$

where σ_{ij}^t is the total stress, and σ_{ij}^p is determined from the yield function via the return mapping algorithm:

$$F^p(\sigma_{ij}^t - \sigma_{ij}^p) = F^p(\sigma_{ij}^t - \Delta\lambda l_{ij}) \tag{3.33}$$

The critical component of this equation is l_{ij}, which is the return direction defined by:

$$l_{ij} = E_{ijkl}\frac{\partial G^p(\sigma_{kl}^t)}{\partial \sigma_{kl}} \tag{3.34}$$

$$\Rightarrow \Delta\varepsilon_{ij}^p = \Delta\lambda\frac{\partial G^p(\sigma_{ij}^t)}{\partial \sigma_{ij}} \tag{3.35}$$

where $G^p(\sigma_{ij})$ denotes the plastic potential function, whose derivative is evaluated at the predictor stress state σ_{ij}^t so as to determine the return direction. The adopted failure surface is the one identified by Menétrey and Willam (1995), whose formulation provides considerable flexibility:

$$F_{3p}^P = \left[\sqrt{1.5}\frac{\rho}{f_c'}\right]^2 + m\left[\frac{\rho}{\sqrt{6}f_c'}r(\theta, e) + \frac{\xi}{\sqrt{3}f_c'}\right] - c = 0 \tag{3.36}$$

where:

$$m = \sqrt{3}\frac{f_c'^2 - f_t'^2}{f_c'f_t'}\frac{e}{e+1} \tag{3.37}$$

$$r(\theta, e) = \frac{4(1 - e^2)\cos^2\theta + (2e - 1)^2}{2(1 - e^2)\cos\theta + (2e - 1)\sqrt{4(1 - e^2)\cos^2\theta + 5e^2 - 4e}} \tag{3.38}$$

m is the cohesion parameter and r an elliptic function; (ξ, ρ, θ) constitute the Haigh-Westergaard coordinates; f_c' and f_t' are the uniaxial compressive and tensile strength, respectively. The curvature of the failure surface is controlled by $e \in \langle 0.5, 1.0\rangle$ (a sharp corner for $e = 0.5$, and circular for $e = 1.0$, see Fig. 3.10. The failure surface position is not fixed but instead can expand and move along the hydrostatic axis (in simulating the hardening and softening stages), as based on the hardening/softening parameter value (κ).

To the contrary, this position can move depending on the magnitude of the strain hardening/softening parameter. Strain hardening is derived from the equivalent plastic strain, which is calculated from $\Delta\varepsilon_{eq}^p = \min(\Delta\varepsilon_{ij}^p)$.

Hardening/softening is controlled by the parameter $c \in \langle 0, 1\rangle$, which evolves during the yielding/crushing process according to:

$$c = \left(\frac{f_c'(\varepsilon_{eq}^p)}{f_c'}\right)^2 \tag{3.39}$$

where $f_c'(\varepsilon_{eq}^p)$ is the hardening/softening law based on a uniaxial test (Fig. 3.11). The law illustrated in Figure 3.11 has an elliptical ascending branch and then a linear post-peak softening branch. The elliptical ascending part depends on the strains:

$$\sigma = f_{c0} + (f_c - f_{c0})\sqrt{1 - \left(\frac{\varepsilon_c - \varepsilon_{sq}^p}{\varepsilon_c}\right)^2} \tag{3.40}$$

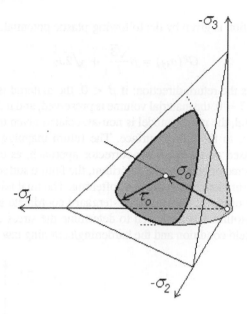

FIGURE 3.10   Failure surface (Menétrey and Willam, 1995)

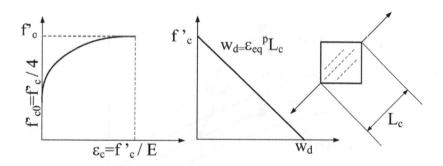

FIGURE 3.11   Compressive hardening and softening, (van Mier, 1986)

while the descending part is based on relative displacements. In order to introduce mesh objectivity, the descending branch is also based on the work of van Mier (1986), whereby the equivalent plastic strain is transformed into displacements through the length scale L_c. This parameter is defined in a manner analogous to the crack band parameter in the fracture model (Fig. 3.9) and corresponds to the projection of element size onto the direction of minimum principal stresses. The square in Eq. 3.39 is due to the quadratic nature of the Menétrey-Willam surface.

The return direction is given by the following plastic potential:

$$G^p(\sigma_{ij}) = \beta\frac{\sqrt{3}}{I_1} + \sqrt{2J_2} \qquad (3.41)$$

where β determines the return direction: if $\beta < 0$, the material is being compacted during crushing; if $\beta = 0$, the material volume is preserved; and if $\beta > 0$, the material is dilating. In general, the plastic model is non-associated given that the plastic flow is not perpendicular to the failure surface. The return mapping algorithm for the plastic model is based on the predictor-corrector approach, as exhibited in Figure 3.12. During the corrector phase of the algorithm, the failure surface moves along the hydrostatic axis to simulate hardening and softening. The final failure surface has its apex located at the origin of the Haigh-Westergaard coordinate system. The secant method-based Algorithm 1 is then used to determine the stress on the surface that satisfies both the yield condition and the hardening/softening law.

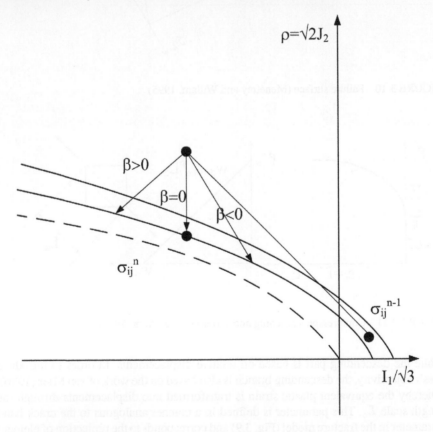

FIGURE 3.12   Plastic Predictor-Corrector Algorithm, (Cervenka and Papanikolaou, 2008)

Algorithm 1: Input: $\sigma_{ij}^{n-1}, \varepsilon_{ij}^{p^{n-1}}, \Delta\varepsilon_{ij}^n$

1. Elastic predictor $\sigma_{ij}^t = \sigma_{ij}^{n-1} + E_{ijkl}\Delta\varepsilon_{kl}^n$
2. Evaluate failure criterion: $f_A^p = F^p(\sigma_{ij}^t, \varepsilon_{ij}^{p^{n-1}}, \Delta\lambda_A = 0$
3. If the failure criterion is violated, i.e. $f_A^p > 0$

   a. Evaluate return direction: $m_{ij} = \dfrac{\partial G^p(\sigma_{ij}^t)}{\partial\sigma_{ij}}$

   b. Return mapping: $F^p(\sigma_{ij}^t - \Delta\lambda_B Em_{ij}, \varepsilon_{ij}^{p^{n-1}}) = 0 \Rightarrow \Delta\lambda_B$

   c. Evaluate failure criterion: $f_B^p = F^p(\sigma_{ij}^t - \Delta\lambda_B Em_{ij}, \varepsilon_{ij}^{p^{n-1}}) + \Delta\lambda_B m_{ij}$

   d. Secant iterations as long as $|\Delta\lambda_A - \Delta\lambda_B| < e$

      i. New plastic multiplier increment: $\Delta\lambda = \Delta\lambda_A - f_A^p \dfrac{\Delta\lambda_B - \Delta\lambda_A}{f_B^p - f_A^p}$

      ii. New return direction: $m_{ij}^{(i)} = \dfrac{\partial G^p(\sigma_{ij}^t - \Delta\lambda Em_{ij}^{(i-1)})}{\partial\sigma_{ij}}$

      iii. Evaluate failure criterion: $f^p = F^p(\sigma_{ij}^t - \Delta\lambda Em_{ij}^{(i)}, \varepsilon_{ij}^p + \Delta\lambda m_{ij}^{(i)})$

      iv. New initial values for secant iterations:

$$* \quad \begin{array}{l} f_B^p < 0 \quad\Rightarrow\quad f_B^p = f^p, \quad \Delta\lambda_B = \Delta\lambda \\ f_B^p \geqslant 0 \quad\Rightarrow\quad f_A^p = f_B^p, \quad \Delta\lambda_A = \Delta\lambda_B, \quad f_B^p = f^p, \quad \Delta\lambda_B = \Delta\lambda \end{array} \tag{3.42}$$

   e. End of secant iteration loop.

4. End of algorithm update of stress and plastic strains. $\varepsilon_{ij}^{p^n} = \varepsilon_{ij}^{p^{n-1}} + \Delta\lambda_B mij^{(i)}$ $chsigma_{ij}^n = \sigma_{ij}^t - \Delta\lambda_B m_{ij}^{(i)}$

3.3.4 COMBINATION OF PLASTICITY AND FRACTURE MODEL

The objective here is to combine the above models into a single model such that plasticity is being used for concrete crushing while the Rankine fracture model applies to cracking. This problem can generally be stated as a simultaneous solution of the two following inequalities.

$$F^p(\sigma_{ij}^{n-1} + E_{ijkl}(\Delta\varepsilon_{kl} - \Delta\varepsilon_{kl}^f - \Delta\varepsilon_{kl}^p)) \leqslant 0 \quad \text{solve for} \quad \Delta\varepsilon_{kl}^p \tag{3.43}$$

$$F^f(\sigma_{ij}^{n-1} + E_{ijkl}(\Delta\varepsilon_{kl} - \Delta\varepsilon_{kl}^p - \Delta\varepsilon_{kl}^f)) \leqslant 0 \quad \text{solve for} \quad \Delta\varepsilon_{kl}^f \tag{3.44}$$

Each inequality depends on the output derived from the other inequality, hence the following iterative scheme is developed.

Algorithm 2:

1. $F^p(\sigma_{ij}^{n-1} + E_{ijkl}(\Delta\varepsilon_{kl} - \Delta\varepsilon_{kl}^{f^{i-1}} + b\Delta\varepsilon_{kl}^{cor(i-1)} - \Delta\varepsilon_{kl}^{p(i)})) \leqslant 0 \text{ solve for} \Delta\varepsilon_{kl}^{p(i)}$
2. $F^p f(\sigma_{ij}^{n-1} + E_{ijkl}(\Delta\varepsilon_{kl} - \Delta\varepsilon_{kl}^{p^{i-1}} - Delta\varepsilon_{kl}^{f(i)})) \leqslant 0 \text{ solve for} \Delta\varepsilon_{kl}^{f(i)}$
3. $\Delta\varepsilon_{ij}^{cor(i)} = \Delta\varepsilon_{ij}^{f(i)} - \Delta\varepsilon_{ij}^{f(i-1)}$
4. The iterative correction of the strain norm between two subsequent iterations can be expressed as: $\|\Delta\varepsilon_{ij}^{cor(i)}\| = (1-b)\alpha^f\alpha^p\|\Delta\varepsilon_{ij}^{cor(i-1)}\|$ where: $\alpha^f = \dfrac{\|\Delta\varepsilon_{ij}^{f(i)} - \Delta\varepsilon_{ij}^{f(i-1)}\|}{\Delta\varepsilon_{ij}^{p(i)}\Delta\varepsilon_{ij}^{p(i-1)}}$ and $\alpha^p = \dfrac{\|\Delta\varepsilon_{ij}^{p(i)} - \Delta\varepsilon_{ij}^{p(i-1)}\|}{\Delta\varepsilon_{ij}^{f(i)}\Delta\varepsilon_{ij}^{f(i-1)}}$

b is an iteration correction or relaxation factor introduced in order to guarantee convergence. Its determination is based on the run-time analysis of α^f and α^p, so as to ensure convergence of the iterative scheme. The parameters α^f and α^p characterize the mapping properties of each model (i.e. plastic and fracture). It is possible to consider each model as an operator mapping the strain increment on the input into a fracture or plastic strain increment on the output. The product of these two mappings must be contractive in order to obtain convergence. The necessary condition for convergence is:

$$|(1 - b)\alpha^f \alpha^p| < 1 \qquad (3.45)$$

If b equals 0, then an iterative algorithm based on recursive substitution is obtained. Convergence can be guaranteed in just two cases:

1. One of the models is not activated (i.e. implies α^f or $\alpha^p = 0$)
2. Neither model exhibits softening and the dilating material is not used in the plastic part, which for the plastic potential in this work would mean $\beta < 0$, (Eq. 3.41). This is a sufficient, though non-necessary, condition to ensure that α^f and $\alpha^p < 1$.

It can be demonstrated that the values of α^f and α^p are directly proportional to the softening rate in each model. Since the softening model typically remains constant whether for a material model or a finite element model, its values do not change significantly between iterations. The scalar b can be selected such that the inequality Eq. 3.45 is always satisfied at the end of each iteration, as based on the current values of α^f and α^p. Three possible scenarios need to be taken into consideration for the appropriate calculation of b:

1. $|\alpha^f \alpha^p| \leqslant \chi$, where χ is related to the requested convergence rate. For a linear rate, it can be set to $\chi = 1/2$, in which case the convergence is satisfactory and $b = -0$.
2. $\chi < |\alpha^f \alpha^p|$, then convergence would be too slow, in which case b can be estimated as $b = 1 - \frac{|\alpha^f \alpha^p|}{\chi}$ in order to increase the convergence rate.
3. $1 \leqslant |\alpha^f \alpha^p|$, then the algorithm is diverging, in which case b should be calculated as $b = 1 - \frac{\chi}{|\alpha^f \alpha^p|}$ to stabilize the iterations.

This approach guarantees convergence as long as the parameters do not change drastically from one iteration to the next, a condition that should be satisfied for smooth and correctly formulated models. The convergence rate depends on material brittleness, dilating parameter β and finite element size. It is advantageous to further stabilize the algorithm by smoothing parameter b during the iterative process:

$$b = \frac{b^{(i)} + b^{(i-1)}}{2} \qquad (3.46)$$

where the superscript i denotes values from two subsequent iterations. This smoothing step will eliminate problems due to the oscillation of correction parameter b. One important condition for convergence of the above Algorithm 2 is that the failure surfaces of both models intersect each other in all possible positions even during hardening or softening. Additional constraints are introduced into the iterative algorithm: if the stress state at the end of the first step violates the Rankine criterion, then the order of the first two steps in Algorithm 2 is reversed. Moreover, in reality, concrete crushing in one direction exerts an effect on cracking in other directions. It is assumed that once the plasticity yield criterion has been violated, the tensile strength in all material directions is set to zero. On the structural level, the secant matrix is used to achieve a robust convergence during the strain localization process. The proposed algorithm for combining plastic and fracture models is shown graphically in Figure 3.13. When

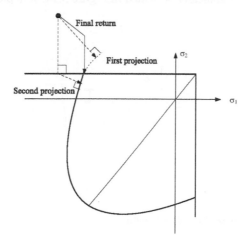

FIGURE 3.13   Schematic description of the iterative process in 2D, (Cervenka and Papanikolaou, 2008)

both surfaces are activated, the behavior is quite similar to the multi-surface plasticity. In contrast with the multi-surface plasticity algorithm, the method proposed herein is more general in the sense that it covers all loading regimes, including physical changes like crack closure. It is currently being developed for just two interacting models, as its extension to multiple models is not a straightforward process.

3.4   NONLINEAR DISCRETE JOINT ELEMENT

This subsection will discuss the nonlinear modeling of concrete using a discrete crack fracture mechanics-based model. It will address two important issues: mixed-mode fracture in homogeneous materials, and interface fracture. A new three-dimensional interface crack model will also be derived through generalizing the classical Hillerborg's fictitious crack model (FCM), which can be recovered if shear displacements are set to zero. Several examples will be used to validate the applicability of this proposed interface crack model.

3.4.1 INTRODUCTION

The classical FCM model defines a relationship between a normal crack opening
and normal cohesive stresses, in assuming no sliding displacements or shear stresses
along the process zone. This assumption is only partially valid for concrete materials.
Based on experimental observation, it is indeed correct to assert that a crack is usually initiated in pure mode I (i.e. opening mode) in concrete, even for mixed-mode
loadings. During propagation however, the crack may curve due to stress redistribution or a non-proportional loading, and significant sliding displacements may develop
along the crack, as schematically presented in Figure 3.14. It is preferable therefore

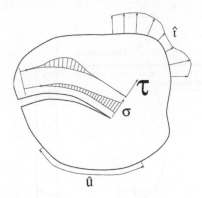

FIGURE 3.14   Mixed-mode crack propagation

to incorporate these shear effects into the proposed crack model.

Finally for concrete dams, it is well accepted that the weakest part of the structure
is the dam-foundation interface, which also happens to be the location of highest
tensile stresses and lowest tensile strength. Given the scope of the present work, this
particular problem must be addressed.

Interface elements were first proposed by Goodman, R.E. and Taylor, R.C. and
Brekke, T.C. (1968) in order to model the nonlinear behavior of rock joints. Since that
time, numerous interface constitutive models have been proposed for a wide range of
applications, such as rock joints (Goodman, R.E. and Taylor, R.C. and Brekke, T.C.,
1968), masonry structures (Lotfi, 1992) and concrete fracture (Stankowski, 1990)
(Feenstra et al., 1991) and (Carol, I. and Bažant, Z.P. and Prat, P.C., 1992).

In the following section, an interface crack model will be proposed and then used to
simulate cracking both in homogeneous concrete and along a rock-concrete interface.
The model presented is a modification of the one first proposed by Carol, I. and Bažant,
Z.P. and Prat, P.C. (1992).

3.4.2 INTERFACE CRACK MODEL

The objective of this section is to develop a physically sound model yet that remains
simple enough for all its parameters to be easily derived from laboratory tests. This

model should be capable of simulating the behavior of rock-concrete and concrete-concrete interfaces.

With this model, the rock-concrete contact is idealized as an interface between two dissimilar materials with zero thickness. The objective therefore is to define relationships between normal and tangential stresses with opening and sliding displacements. The notation used for such an interface model is illustrated in Figure 3.15. The major

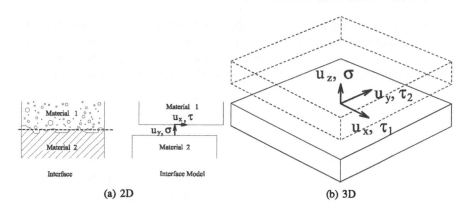

(a) 2D   (b) 3D

FIGURE 3.15   Interface idealization and notations

premises underlying this model are:

1. Shear strength depends on the normal stress.
2. Softening is present in both shear and tension.
3. A residual shear strength exists due to friction along the interface, which in turn depends on the normal compressive stress.
4. The reduction in strength, i.e. softening, is caused by crack formation.
5. When the interface is totally destroyed, both the normal and shear stiffness values equal zero.
6. Under normal compressive stresses, neither the shear nor the normal stiffness decreases to zero. In addition, should a compressive stress be introduced in the normal direction following a full crack opening, two faces of the interface would come into contact, and both the tangential and normal stiffness would become nonzero.
7. Irreversible relative displacements are caused by broken segments of the interface material as well as by friction between the two crack surfaces.
8. Interface roughness causes opening displacements (i.e. dilatancy) when subjected to sliding displacements.
9. Dilatancy vanishes with increasing sliding or opening displacements.

In the proposed model, the strength of an interface is described by the following failure function:

$$F = (\tau_1^2 + \tau_2^2) - 2c \tan(\phi_f)(\sigma_t - \sigma) - \tan^2(\phi_f)(\sigma^2 - \sigma_t^2) = 0 \qquad (3.47)$$

where:

- c is the cohesion.
- ϕ_f is the angle of friction.
- σ_t is the tensile strength of the interface.
- τ_1 and τ_2 are the two tangential components of the interface traction vector.
- σ is the normal traction component.

The shape of the failure function in the two-dimensional case is shown in Figure 3.16, which corresponds to the failure criterion first proposed by Carol, I. and Bažant, Z.P. and Prat, P.C. (1992). The general three-dimensional failure function is obtained by simple rotation around the σ-axis. The evolution of the failure function

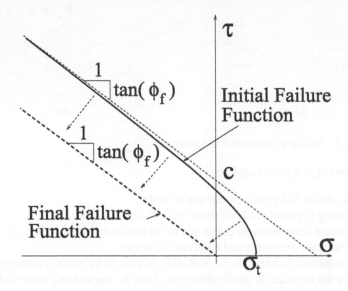

FIGURE 3.16   Failure function

is based on a softening parameter u^{ieff} that expresses the norm of the inelastic displacement vector u^i. This inelastic displacement vector is obtained by decomposing displacement vector u into an elastic part u^e and an inelastic part u^i. The inelastic part can subsequently be broken down into plastic (i.e. irreversible) displacements u^p and fracturing displacements u^f. Plastic displacements are assumed to be caused by friction between crack surfaces, while fracturing displacements are assumed to result from the formation of micro-cracks.

$$F = F(c, \sigma_t, \phi_f), \quad c = c(u^{ieff}), \quad \sigma_t = \sigma_t(u^{ieff})$$
$$u = u^e + u^i, \quad u^i = u^p + u^f$$
$$u^{ieff} = ||u^i|| = (u_x^{i\,2} + u_y^{i\,2} + u_z^{i\,2})^{1/2}$$

(3.48)

In this work, both linear and bilinear relationships have been used for $c(u^{\text{ieff}})$ and $\sigma_t(u^{\text{ieff}})$.

$$
\left.
\begin{aligned}
c(u^{\text{ieff}}) &= c_0\left(1 - \frac{u^{\text{ieff}}}{w_c}\right) & \forall\, u^{\text{ieff}} < w_c \\
c(u^{\text{ieff}}) &= 0 & \forall\, u^{\text{ieff}} \geq w_c \\
w_c &= \frac{2G_F^{IIa}}{c_0}
\end{aligned}
\right\} \text{ linear for cohesion}
$$

$$
\left.
\begin{aligned}
c(u^{\text{ieff}}) &= c_0 + u^{\text{ieff}}\frac{s_{1c}-c_0}{w_{1c}} & \forall\, u^{\text{ieff}} < w_{1c} \\
c(u^{\text{ieff}}) &= s_c\left(1 - \frac{u^{\text{ieff}}-w_{1c}}{w_c-w_{1c}}\right) & \forall\, u^{\text{ieff}} \in \langle w_{1c}, w_c\rangle \\
c(u^{\text{ieff}}) &= 0 & \forall\, u^{\text{ieff}} > w_c \\
w_c &= \frac{2G_F^{IIa}-c_0 w_{1c}}{s_{1c}}
\end{aligned}
\right\} \text{ bilinear for cohesion}
$$

$$(3.49)$$

$$
\left.
\begin{aligned}
\sigma_t(u^{\text{ieff}}) &= \sigma_{t0}\left(1 - \frac{u^{\text{ieff}}}{w_\sigma}\right) & \forall\, u^{\text{ieff}} < w_\sigma \\
\sigma_t(u^{\text{ieff}}) &= 0 & \forall\, u^{\text{ieff}} \geq w_{\sigma_t} \\
w_\sigma &= \frac{2G_F^I}{\sigma_{t0}}
\end{aligned}
\right\} \text{ linear for tensile strength}
$$

$$
\left.
\begin{aligned}
\sigma_t(u^{\text{ieff}}) &= \sigma_{t0} + u^{\text{ieff}}\frac{s_{1\sigma}-\sigma_{t0}}{w_{1\sigma}} & \forall\, u^{\text{ieff}} < w_{1\sigma} \\
\sigma_t(u^{\text{ieff}}) &= s_{1\sigma}\left(1 - \frac{u^{\text{ieff}}-w_{1\sigma}}{w_{\sigma_t}-w_{1\sigma}}\right) & \forall\, u^{\text{ieff}} \in \langle w_{1\sigma}, w_\sigma\rangle \\
\sigma_t(u^{\text{ieff}}) &= 0 & \forall\, u^{\text{ieff}} > w_\sigma \\
w_\sigma &= \frac{2G_F^I-\sigma_{t0} w_{1\sigma}}{s_{1\sigma}}
\end{aligned}
\right\} \begin{array}{l} \text{bi-linear for} \\ \text{tensile strength} \end{array}
$$

$$(3.50)$$

where G_F^I and G_F^{IIa} are mode I and II fracture energies. s_{1c}, w_{1c} and $s_{1\sigma}, w_{1\sigma}$ are the coordinates of the breakpoint in the bilinear softening laws for cohesion and tensile strength respectively. The critical opening and sliding, corresponding to zero cohesion and tensile strength, are denoted by w_σ and w_c. Their determination is based on the condition that the area under the linear or bilinear softening law must be equal to G_F^I and G_F^{IIa} respectively. The significance of these symbols can best be explained through Figure 3.17. It should be noted that G_F^{IIa} is not the pure mode II fracture

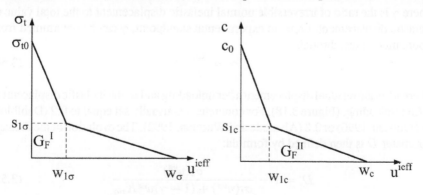

FIGURE 3.17   Bilinear softening laws

energy (i.e. the area under a τ-u_x curve), but rather the energy dissipated during a shear test with high normal confining stress. This parameter was first introduced by Carol, I. and Bažant, Z.P. and Prat, P.C. (1992) in their microplane model; the ensuing representation seems to be more favorable to the pure mode II fracture energy G_F^{II}. The determination of G_F^{II} would require a pure shear test without confinement, which is extremely difficult to perform. Alternatively, a G_F^{IIa} test requires a substantial normal confinement and therefore is easier to accomplish. Furthermore, if G_F^{II} is used, the entire shear-compression region of the interface model would be an extrapolation of observed behavior, whereas the second approach represents an interpolation between the upper bound G_F^{IIa} and lower bound G_F^I. The residual shear strength is obtained from the failure function by setting both c and σ_t equal to 0, which corresponds to the final shape of the failure function in Figure 3.16 and is given by:

$$\tau_1^2 + \tau_2^2 = \tan^2(\phi_f)\,\sigma^2 \tag{3.51}$$

Stiffness degradation is modeled through a damage parameter, $D \in \langle 0, 1 \rangle$, which is a relative measure of the fractured surface. Thus, D is related to the secant of the normal stiffness K_{ns} in the uniaxial case:

$$D = \frac{A_f}{A_o} = 1 - \frac{K_{ns}}{K_{no}} \tag{3.52}$$

where K_{no} is the initial normal stiffness of the interface, and A_o and A_f are the total interface area and fractured area, respectively. It is assumed that the damage parameter D can be determined by transforming the mixed-mode problem into an equivalent uniaxial problem (Figure 3.18). In this equivalent uniaxial problem, the normal inelastic displacement is set equal to u^{ieff}; then, the secant normal stiffness can be determined from:

$$K_{ns} = \frac{\sigma}{u - u^p} = \frac{\sigma_t(u^{ieff})}{u^e + u^p + u^f - u^p} = \frac{\sigma_t(u^{ieff})}{\sigma_t(u^{ieff})/K_{no} + (1 - \gamma)u^{ieff}} \tag{3.53}$$

where γ is the ratio of irreversible normal inelastic displacement to the total value of inelastic displacement. From an experimental standpoint, γ can be determined from a pure mode I test through:

$$\gamma = \frac{u_p}{u_i} \tag{3.54}$$

where u^p is the residual displacement after unloading and u^i the inelastic displacement before unloading. (Figure 3.18). For concrete, γ is usually set equal to 0.2 (Dahlblom and Ottosen, 1990) or 0.3 (Alvaredo and Wittman, 1992). The evolution of the damage parameter D is then defined by formula:

$$D = 1 - \frac{\sigma_t(u^{ieff})}{\sigma_t(u^{ieff}) + (1 - \gamma)u^{ieff}K_{no}} \tag{3.55}$$

which is obtained by substituting Equation 3.53 into Eq. 3.52.

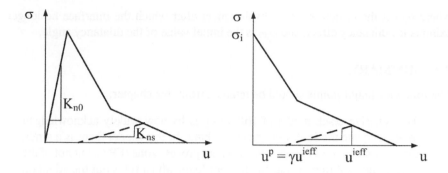

FIGURE 3.18   Stiffness degradation in the equivalent uniaxial case

The stress-displacement relationship of the interface is expressed as:

$$\boldsymbol{\sigma} = \alpha \mathbf{E}(\boldsymbol{u} - \boldsymbol{u}^p) \tag{3.56}$$

where: (a) $\boldsymbol{\sigma}$ is the vector of tangential and normal stress at the interface.

$$\boldsymbol{\sigma} = \{\tau_1, \tau_2, \sigma\}^T \tag{3.57}$$

(b) α is the integrity parameter defining the relative active area of the interface and is correlated with damage parameter D.

$$\alpha = 1 - \frac{|\sigma| + \sigma}{2|\sigma|} D \tag{3.58}$$

Let's note that α can be different from 1 only if the normal stress σ is positive (i.e. the interface is in tension). In other words, damage parameter D is activated only if the interface is in tension. In compression, the crack is assumed to be closed, with full contact between the two crack surfaces. The activation of D is controlled through the fraction $\frac{|\sigma| + \sigma}{2|\sigma|}$, which equals one if σ is positive and zero otherwise.

(c) $\mathbf{E}$ is the elastic stiffness matrix of the interface.

$$\mathbf{E} = \begin{bmatrix} K_{to} & 0 & 0 \\ 0 & K_{to} & 0 \\ 0 & 0 & K_{no} \end{bmatrix} \tag{3.59}$$

It should be noted here that the off-diagonal terms in the elastic stiffness matrix $\mathbf{E}$ of the interface are all equal to zero, which implies that no dilatancy is considered in the elastic range. Dilatancy is introduced later after reaching the failure limit through the iterative solution process. The dilatancy of the interface is given by a dilatancy angle ϕ_d, which once again is assumed to be a function of u^{ieff}. In the proposed model, a linear relationship has been assumed therein:

$$\begin{aligned} \phi_d(u^{\text{ieff}}) &= \phi_{d0}(1 - \frac{u^{\text{ieff}}}{u_{dil}}) & \forall u^{\text{ieff}} \leq u_{dil} \\ \phi_d(u^{\text{ieff}}) &= 0 & \forall u^{\text{ieff}} > u_{dil} \end{aligned} \tag{3.60}$$

where u_{dil} is the critical relative displacement after which the interface no longer exhibits the dilatancy effect, and ϕ_{d0} is the initial value of the dilatancy angle.

3.5 SUMMARY

The following major points should be retained from this chapter:

1. The cohesive crack model of Hillerborg is by now widely acknowledged as an adequate model for concrete cracking. In this model there is a strain discontinuity across the so-called fracture process zone (FPZ), but not of the stress. The FPZ in concrete can be very long, which rules out the adoption of linear elastic fracture mechanics.
2. AAR's swelling will almost always result in cracking. Cracking may be extensive and non-structural, in which case one should adopt the smeared crack model. The restrained AAR expansion on the other hand may result in localized structural crack(s). In this case the discrete crack model should be adopted.

4 Validation

Chapter written with Antoine Tixier

Before any finite element code is used, it must be validated. Unfortunately, this basic elementary step is too often forgotten by engineers who (conveniently) ignore this most basic step.

This chapter will first address the role of benchmarks, and then present results of Benchmark problems, devised by the author and Prof. Sellier (Appendix A) obtained by the Author's model (and coded in the MERLIN finite element code).

4.0.1 BENCHMARKS

A number of benchmark problems have been proposed in the past. To the best of the author's knowledge, these problems have focused exclusively on existing dams addressed within the ICOLD (International Commission on Large Dams) International Benchmark Workshops on Numerical Analysis of Dams. The sixth (Theme-A, Salzburg) and eighth (Wuhan) benchmark workshops invited participants to analyze the Pian Telessio and Poglia Dams, respectively. No study was submitted for the former and only two were received for the latter. The most recent benchmark study has been centered on the Kariba Dam. Given that a benchmark study seeks to compare and contrast models at the level of their various components by analyzing the same problem, benchmarking through a dam renders such an exercise completely infeasible due to the great number of variables involved simultaneously. By way of comparison, during the MECA project (organized by the *Electricité de France* Research and Development Department), 13 teams sought to analyze four problems that were representative of the difficulties encountered when modeling concrete. The attempt therein was to identify a material model for concrete suitable for implementation in a general-purpose finite element code for the purpose of analyzing the safety of large-scale concrete structures, such as cooling towers, dams or nuclear reactor vessels, (de Borst, 2003).

More recently, RILEM TC 219-ACS (Alkali-Aggregate Reactions in Concrete Structures) subcommittee M (Modeling of Structures) has been examining the development of *practical guidance on the use of numerical models to reassess AAR affected structures*, (Seignol and Godart, 2012). Yet, so far this committee has not established a short list of benchmark tests to evaluate the proposed model's capabilities. The working group concluded that "building all these tests would represent a large amount of work, maybe even exceeding the team's capabilities", (Seignol, 2010).

In light of the above considerations, Saouma and Sellier (2010) submitted to the engineering community a series of carefully selected problems requiring just one parameter at a time and ultimately leading to the analysis of a synthetic dam model. Such a gradual and methodological approach is the only means by which computational models can be compared, contrasted and assessed.

The benchmark statement, see Appendix A, has been broken down as follows:

P0 brief description of the finite element model used.

Materials this first part focuses on the ability of the model to capture the following variables (under free expansion for the first five cases)

P1 Calibration and prediction of constitutive models without any reference to AAR.

P2 Drying and shrinkage.

P3 Creep.

P4 Temperature.

P5 Relative humidity.

P6 Effect of confinement.

Structural Response under complex states

P7 Effect of Internal Reinforcement

P8 Idealized Dam subjected to temperature, stress, RH, creep and crack closure.

P0 is comprehensively addressed in Chapter 3, and the following section will present all other results obtained with the Merlin finite element code (Saouma et al., 2010) applied to the author's model.

4.1 BENCHMARK RESULTS

4.1.1 P1: CONSTITUTIVE MODEL

As previously stated, Merlin's constitutive model is completely distinct from the AAR model and will be tested first in this section. P0 therefore seeks to capture the nonlinear response of concrete when subjected to a load history covering both tension and compression. A simulation has been run for the 16 x 32-cm cylinder shown in Figure 4.1 (this same mesh will be used for all test problems).

Two simulations have been conducted, the first without AAR and for a strain history given by

$$0 => 1.5\frac{f'_t}{E} => 0 => 3\frac{f'_t}{E} => 1.5\epsilon_c => 0 => 1.5\epsilon_c \qquad (4.1)$$

and the second for an identical strain history yet preceded by an AAR expansion. For both cases, the properties are listed in Table 4.1. Figure 4.2 plots the load-displacement curve at the top of the cylinder. In both cases, the load-displacement curve at the top of the cylinder surface is being plotted. The AAR expansion vs. time is also plotted. Let's begin by observing the model's nonlinear response with a peak compressive strength of approx. -38 MPa and the onset of nonlinearity of about -13 MPa. The tensile strength of 3.5 MPa is also reduced by the specified $\beta_t = 0.4$ to about 1.4 MPa; moreover, the elastic modulus degradation of β_E has also been clearly captured.

TABLE 4.1

Constitutive Model - AAR Properties

Material parameters		Unit	Value
Young's Modulus (Static)	E_s	MPa	37,300.
Poisson's Ratio (Static)	ν_s	-	0.2
Tensile Strength	f_t	MPa	3.5
Fracture Energy	Γ_f	MN/m	$1.00\ 10^{-4}$
Compressive Strength	f_c	MPa	-38.4
Compressive Critical Displacement	w_d	m	-0.5
Return Direction Factor	β	-	0.
Failure Surface Roundness Factor	e	-	0.52
Onset of Compression Nonlinearity	f_{c0}	MPa	-13.3
Plastic Strain at f_c	ϵ_c	-	$-2.0\ 10^{-3}$
AAR parameters		Unit	Value
Maximum Volumetric AAR Strain	ϵ^∞	-	50×10^{-4}
Characteristic Time	τ_{car}	ATU	3.34
Latency Time	τ_{lat}	ATU	8.29
Activation Energy for Characteristic Time	U_C	K	5,400.
Activation Energy for Latency Time	U_L	K	9,400.
Reference Test Temperature	T_0	$^\circ C$	38.
Tensile Strength	f_t	MPa	3.5
Compressive Strength	f_c	MPa	-38.4
Residual Reduction Factor in Tension	Γ_r	-	0.1
Fraction of Tension pre-AAR Comp. Red.	γ_t	-	0.23
Upper Compressive Stress Limit	σ_u	MPa	-10.
Shape Factor for Γ_c	β	-	-2.
Reduction Factor for Young's Modulus	β_E	-	0.35
Reduction Factor for Tensile Strength	β_f	-	0.4

4.1.2 P2: DRYING AND SHRINKAGE

This problem has not been addressed for two reasons. First, the importance of drying and shrinkage is really quite minimal for dams (except in the case of an insignificant layer of concrete), hence the overall impact of such a simulation is almost nil. Furthermore, the effect of shrinkage is mitigated by the difference in time scales for shrinkage and AAR. This difference however is only relevant for massive hydrostatic facilities and not for reinforced concrete structures exposed to air under varying humidity conditions (though an accurate structural analysis aimed at assessing the effect of AAR for such structures is seldom, if ever, warranted). Second, the Merlin code in its current form is unable to address shrinkage.

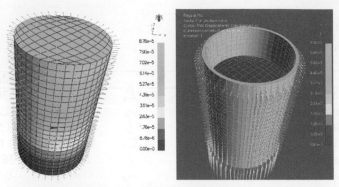

(a) Finite element mesh of a concrete   (b) Finite element mesh of a concrete
cylinder   cylinder and steel jacket

FIGURE 4.1   Finite element model for concrete cylinders

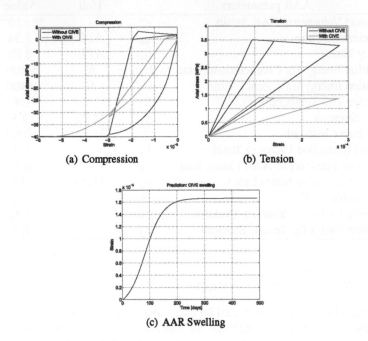

(a) Compression   (b) Tension

(c) AAR Swelling

FIGURE 4.2   Constitutive Model - Results

4.1.3   P3: CREEP

It is now well established that creep plays a dominant role in the long-term response of
massive concrete structures subjected to even moderate compression. In some cases,
AAR strain may indeed be offset by the creep-induced contraction.

One simple way to account for creep is to determine the time-varying creep coefficient ($\Phi(t)$), as follows:

$$\sigma(t) = \frac{E_0}{1 + \phi}\varepsilon(t) \Rightarrow \phi(t) = \frac{E_0\epsilon(t)}{\sigma(t)} - 1 \qquad (4.2)$$

Since creep cannot be explicitly modeled in Merlin's code, Young's modulus is modified at each time step according to:

$$E(t) = \frac{E_0}{1 + \phi(t)} \qquad (4.3)$$

Fig. 4.3(a) shows the experimentally determined vertical strain *vs* time (for non-reactive concrete), as reported by Multon (2004), and the corresponding computed $\phi(t)$ in Fig. 4.3(b). A linear (visco)elastic constitutive model and the AAR parameters

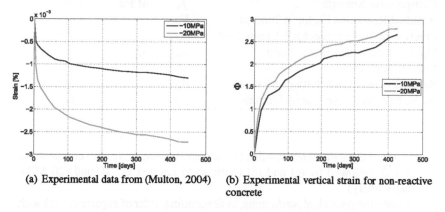

(a) Experimental data from (Multon, 2004)   (b) Experimental vertical strain for non-reactive concrete

FIGURE 4.3   Determination of creep coefficient Φ from non-reactive concrete

shown in Table 4.2 are used for a 13 x 24-cm cylindrical concrete specimen.

Nine numerical simulations of experimental tests have been performed:

Validation AAR expansion at -2, -5, -10 and -20 MPa both with (using the creep coefficients shown in Fig. 4.4) and without creep.

Prediction Based on a time-varying compressive stress history.

where traction was applied on the frictionless top of the cylinder. Convergence criteria (energy error, relative residual error, absolute residual error and displacement error) were all set to 0.01.

In Fig. 4.4(a), we examine both numerical and experimental axial strain:

- In the absence of creep, the experimental (a) and numerical (b) results without creep are reasonably similar, while in the absence of an axial compressive stress they are higher.

TABLE 4.2

Effect of Creep - Material and AAR Properties

Material parameters		Unit	Value
Young's Modulus	E_s	MPa	33600.
Poisson's Ratio	ν_s	-	0.2
Maximum Volumetric AAR Strain	ϵ^∞	-	50×10^{-4}
Characteristic Time	τ_{car}	ATU	3.34
Latency Time	τ_{lat}	ATU	8.29
Activation Energy for Characteristic Time	U_C	K	5,400.
Activation Energy for Latency Time	U_L	K	9,400.
Reference Test Temperature	T_0	$°C$	38.
Tensile Strength	f_t	MPa	10.
Compressive Strength	f_c	MPa	-30.
Residual Reduction Factor in Tension	Γ_r	-	0.1
Fraction of Tension pre-AAR Comp. Red.	γ_t	-	0.23
Upper Compressive Stress Limit	σ_u	MPa	-10.
Shape Factor for Γ_c	β	-	-2.
Reduction Factor for Young's Modulus	β_E	-	1.
Reduction Factor for Tensile Strength	β_f	-	1.

- Among the experimental results, the largest swelling is: (a) (no compressive stress), followed by (c) and (e) (with axial stresses of -10 and -20 MPa, respectively).
- Among the numerical predictions, in descending order of expansion: (b) with no axial stresses, followed by (c), (d), and (f), where the corresponding axial stresses are 0, -10 and -20 MPa, respectively.

In Fig. 4.4(b), the -2 MPa stress is still too low to overcome the AAR expansion, thus making it the only case where a positive strain is taking place. For stresses above -10 MPa, the AAR equals zero, and the combined elastic and AAR strains are well into the negative range, while for -5 and -10 MPa, the net axial strain is nearly zero.

In Fig. 4.4(c), we examine radial strain. In this axisymmetric problem, let's note that with an imposed axial stress of -10 and -20 MPa, both experimental and numerical strains are roughly equal to 2.5×10^{-3}, which is half ε^∞, thus reinforcing the notion of an AAR strain redistribution (or anisotropic expansion), as observed by experimenters as well as the author's model. Therefore, the smaller the imposed axial stress, the smaller the final AAR radial strain (i.e. equal to approx. ε^∞ in the absence of creep).

Fig. 4.4(d) focuses on the radial strain in this given axisymmetric problem. Let's note that both the experimental and numerical strain values are roughly equal to 2.5×10^{-3}, which is half ε^∞, thus once again reinforcing the notion of an AAR strain redistribution (or anisotropic expansion), as observed by researchers and embedded

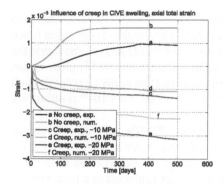

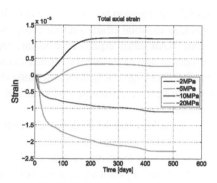

(a) Influence of creep in AAR swelling, total axial strain

(b) Total axial strain (numerical)

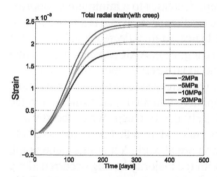

(c) Total radial strain with creep (numerical)

(d) Influence of creep in AAR swelling, total radial strain

FIGURE 4.4   Numerical results of calibration for creep - Part 1

into the author's model. It should also be pointed out that the radial strain is mildly influenced by the Poisson's radial expansion.

The next step consists of evaluating each stress value separately.

In Fig. 4.5(a), the -2 MPa value indicates that the effect of creep is nearly nonexistent. The largest expansion is radial (between one-third and half of ε^{∞}, while the lowest is also radial (less than one-third of ε^{∞}). Creep exerts no influence either on axial strain, which remains positive (i.e. the expansion is greater than the elastic/creep contraction), or on radial strain, which also remains positive (i.e. expansion is greater than Poisson's effect due to the elastic/creep contraction).

In Fig. 4.5(b), the stress has now become -5 MPa, and observations are the same as in the previous case of -2 MPa; however, this greater imposed stress has accentuated the effect.

Figs. 4.5(c) and 4.5(d) correspond to an axial strain less than -10 MPa, (c) with axial strains and (d) with radial strains. Curves (a) and (b) in Fig. 4.5(c) reveal that creep does not exert any impact on vertical strain, which is a correct interpretation:

no AAR expansion is expected in this direction if stress exceeds -10 MPa. Curve (c) provides the total strain without creep, which is equivalent to the elastic strain (once again, AAR strain equals zero in this case). Lastly, curves (d) and (e) show that the numerical and experimental strains in the axial direction are reasonably close to one another. For Fig. 4.5(d), let's note that creep does not exert any impact, except through Poisson's effect. Also note that a value equal to half of ε^{∞} i.e. 2.5×10^{-5} is ultimately reached.

Figs. 4.5(e) and 4.5(f) depict the axial strain under -20 MPa, (e) with axial strains and (f) with radial strains. The same conclusions can be drawn as with -10 MPa, i.e. creep does not exert any impact on AAR strains, except through Poisson's effect.

Moreover, a prediction for the response of a cylinder subjected to a time-varying axial stress, as shown in Fig. 4.6(a), has been derived. Using an average of the two $\phi(t)$ (corresponding to -10 and -20 MPa), this response is given in Fig. 4.6. For starters, the vertical elastic strain, compounded by creep, decreases to a minimum of approx. -2.5×10^{-4} at about 20 days. At this point, the AAR expansion rate is almost nil, and certainly less than the contraction due to creep; hence, strain is decreasing. Next, AAR begins to expand and strain is once again increasing; here, AAR expansion is starting to overcome the elastic strain. Afterwards, the axial stress is increased from -5 to -10 MPa, while the elastic strain compounded with creep causes further contraction. At -10 MPa, the AAR axial expansion is completely inhibited (and redirected in the radial direction), and the strain increase is entirely due to creep. When the stress is again lowered from -10 to -5 MPa, a rebound occurs and from that point forward both creep contraction and (reduced) AAR expansion are at work. Yet at this same point, the propensity for AAR has been exhausted, with the majority having occurred along the radial direction (which is close to 2.5×10^{-3} at the end), meaning that the elastic strain cannot be compensated. In the radial direction however, the opposite behavior is observed, as a result of: the basic AAR expansion in this direction, the redirected AAR expansion between 100 and 300 days, and Poisson's effect.

4.1.4 P4: EFFECT OF TEMPERATURE

All chemical reactions are thermodynamically driven. Reactive concrete expansions vary widely with the temperature ranges typically encountered in the laboratory. It is of paramount importance therefore that the reaction kinetics are capable of capturing this dependence.

A simulation of Larive's tests (Larive, 1998a) on a 13 x 24-cm concrete cylinder is thus performed using the corresponding properties given in Table 4.3.

Three simulations have actually been performed:

Validation : By simulating the free expansion at $23^{o}C$ and $38^{o}C$, whose experimental data are shown in Fig. 4.7(a), the large variability should be noticeable.

Prediction for a harmonic temperature variation given by

$$T(days) = \frac{T_{max} - T_{min}}{2} \sin(2\pi \frac{t/7 - 16}{52}) + \frac{T_{max} + T_{min}}{2} \qquad (4.4)$$

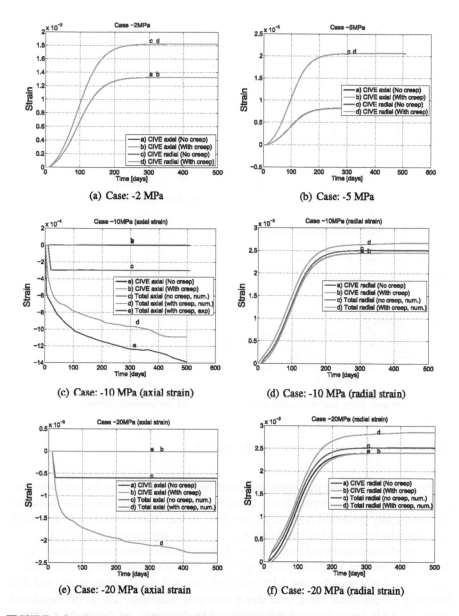

(a) Case: -2 MPa

(b) Case: -5 MPa

(c) Case: -10 MPa (axial strain)

(d) Case: -10 MPa (radial strain)

(e) Case: -20 MPa (axial strain

(f) Case: -20 MPa (radial strain)

FIGURE 4.5   Numerical results of calibration for creep - Part 2

where T_{max} and T_{min} are 25^0C and 0^oC, respectively, Fig. A.9.

Once again, the analyses have been performed on a cylinder and all convergence criteria (Energy Error, Relative Residual Error, Absolute Residual Error and Displacement Error) were set to 0.01.

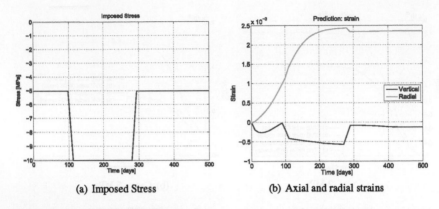

(a) Imposed Stress   (b) Axial and radial strains

FIGURE 4.6   Predicted numerical results

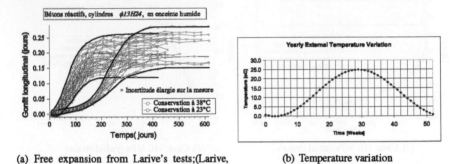

(a) Free expansion from Larive's tests;(Larive,   (b) Temperature variation
1998a)

FIGURE 4.7   Data used for calibration and prediction

Fig. 4.8(a) shows the predicted expansion vs. time for the two temperatures. As anticipated, expansion occurs much faster at the higher temperature. Furthermore, the numerically predicted AAR strain lies close to the value derived experimentally, based on which critical data were calibrated (ε^{∞}, τ_{lat} and $\tau char$), Fig. 4.7(a).

The prediction results are shown in Fig. 4.8(b). The harmonic thermal strain ($\epsilon_{th} = \alpha.\Delta T$) is given first, while the strain caused by AAR is simply expressed as $\epsilon_{AAR} = \epsilon_{total} - \epsilon_{th}$. Let's note that the AAR strain remains flat at low temperature (and thus plateaus); on the other hand, total strain increases with time due to the combined effects of AAR expansion and temperature. The decreases are driven by a drop in temperature.

4.1.5   P5: RELATIVE HUMIDITY

Relative humidity plays a critical role in the expansion of AAR-affected concrete. It is now well established (Poole, 1992) that AAR will start to expand at an RH equal to at least 80% and, afterwards, will continue to increase with RH. Capra and Bournazel

TABLE 4.3

Effect of Temperature - Material and AAR Properties

Material parameters		Unit	Value
Thickness	h	m	1.
Mass Density	ρ	Gg/m^3	$2.4\ 10^{-3}$
Coefficient of Thermal Expansion	α	$/^\circ C$	$1.0\ 10^{-5}$
Young's Modulus	E_s	MPa	33600.
Poisson's Ratio (Static)	ν_s	-	0.2
Tensile Strength	f_t	MPa	10.
Compressive Strength	f_c	MPa	-30.
AAR parameters		Unit	Value
Maximum Volumetric AAR Strain*	ϵ^∞	-	65.4×10^{-4}
Characteristic Time*	τ_{car}	ATU	3.34
Latency Time*	τ_{lat}	ATU	8.29
Activation Energy for Characteristic Time*	U_C	K	5,400.
Activation Energy for Latency Time*	U_L	K	9,400.
Reference Test Temperature	T_0	$^\circ C$	38.
Tensile Strength	f_t	MPa	10.
Compressive Strength	f_c	MPa	-30.
Residual Reduction Factor in Tension	Γ_r	-	0.1
Fraction of Tension pre-AAR Comp. Red.	γ_t	-	0.23
Upper Compressive Stress Limit	σ_u	MPa	-10.
Shape Factor for Γ_c	β	-	-2.
Reduction Factor for Young's Modulus	β_E	-	1.
Reduction Factor for Tensile Strength	β_f	-	1.

* Data extracted from Larive's tests.

(1998) suggested the following simple equation:

$$\varepsilon^{RH} = RH^m \varepsilon^{100\%} \tag{4.5}$$

where m is determined to equal 8 through a regression analysis of experimental data. Again, three simulations have been performed:

Calibration Two analyses with an external relative humidity of 100 and 30 percent using the experimental dataset of Multon (2004), as shown in Figs. A.10 and A.11, respectively.

Prediction of the expansion for a cylinder subjected to the RH histogram shown in Fig. 4.10

The impact of relative humidity on AAR swelling has been modeled as a modification of the final volumetric AAR strain according to the equation: $\epsilon^{AAR}(t) =$

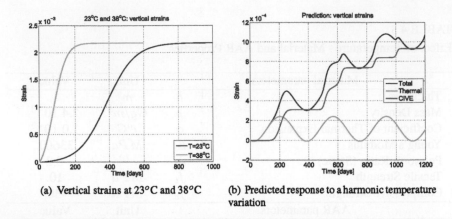

(a) Vertical strains at 23°C and 38°C   (b) Predicted response to a harmonic temperature variation

FIGURE 4.8   Numerical results for the calibration and prediction of temperature effects

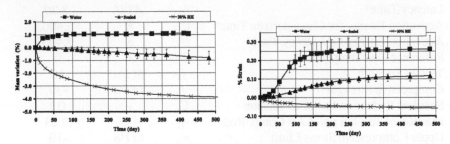

(a) Mass variation for reactive concrete under various RH conditions   (b) Strain variation for reactive concrete under various RH conditions

FIGURE 4.9   Mass and strain variations under various RH conditions;Multon (2004)

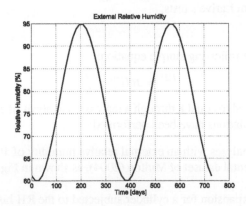

FIGURE 4.10   Cyclic variation of relative humidity used to simulate the prediction

$\epsilon_\infty^{AAR}.RH^8(t)$ A constant relative humidity of 80% will thus lead to multiplying the final volumetric AAR strain by 0.17.

The factor 8 however has not yet been well established, which led to simulating the Prediction case using three different values of this factor: 7, 8 and 9.

All convergence criteria (Energy Error, Relative Residual Error, Absolute Residual Error and Displacement Error) have been set equal to 0.01.

The results of the first simulation are shown in Fig. 4.11(a) for expansion in terms of RH; as expected at 30% RH, expansion is practically nil. Fig. 4.11(b) shows the effect of factor m on this expansion: the higher the exponent value, the lower the level of expansion. Let's point out that in this case, no attempt was made to calibrate input data with experimental results, and the final AAR-induced strain was set to 0.5% just like in most of the previous simulations.

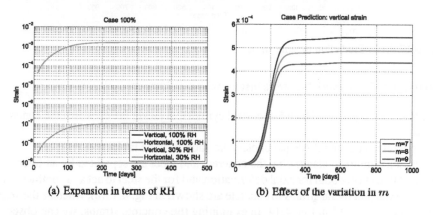

(a) Expansion in terms of RH   (b) Effect of the variation in m

FIGURE 4.11   Calibration: Effect of Relative Humidity

A simulation with a cyclic variation of relative humidity, see Fig. 4.10, will be performed next; these results indicate a sensitivity to m (Fig. 4.12). As mentioned above, the higher m value leads to reduced expansion.

4.1.6   P6: CONFINEMENT

It has long been recognized that confinement inhibits reactive concrete expansion, (Charlwood et al., 1992), (Léger et al., 1996) and, more recently, Multon and Toutlemonde (2006). This test series has sought to ensure that such inhibition is properly captured by the numerical model.

With respect to confinement, five simulations have been performed:

Calibration Based on Multon's thesis, according to which four cases are to be considered:
- a) Free expansion, no confinement
- b) -10 MPa vertical stress, no confinement
- c) Free expansion, confinement
- d) -10 MPa vertical stress, confinement

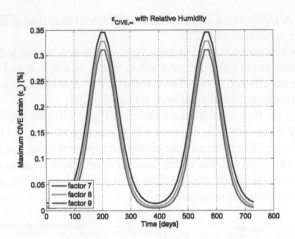

FIGURE 4.12   Prediction in terms of m

Prediction for the variable stress history with confinement

Confinement is provided by a 5-mm steel cylindrical jacket inside which the concrete is cast.

Concrete, steel and AAR parameters are shown in the tables 4.4.

From a numerical standpoint, the concrete and steel are separated by an interface element that allows both for axial deformation and for the steel to act as a confinement.

The axial and radial strains in concrete are shown in Fig. 4.13(a), whereas the steel strains are displayed in Fig. 4.14. In examining the concrete strains, we can observe the following (in descending strain order):

1. (b) Radial, -10 MPa, no confinement. Since AAR is inhibited in the axial direction by the -10 MPa axial stress, it is entirely redirected in the radial direction. The AAR strain is about 15×10^{-4}, corresponding to a total volumetric AAR strain of 30×10^{-4}, which is approximately equal to the specified $\varepsilon^{\infty} = 28.8 \times 10^{-4}$.

2. (c) Axial, free with confinement. This corresponds to the radial expansion inhibiting overall expansion and is hence redirected in the free axial direction. As expected, the magnitude is roughly twice as high.

3. (a) Axial and radial, free no confinement. These are both equal to approximately 9×10^{-4} at 350 days, which is indeed equal to one-third of the specified $\varepsilon^{\infty} = 0.288\%$ in the presence of unconstrained isotropic expansion.

4. (d) -10 MPa confinement, radial. Following an initial increase due to Poisson's effect, some swelling occurs but is being partially inhibited.

5. (c) Radial, free confinement. Unlike the previous case, no initial strain is present and swelling is gradually increasing. Swelling is reduced since most of it is occurring in the axial direction.

TABLE 4.4

Effect of Confinement - Material and AAR Properties

Concrete		Unit	Value
Mass Density	ρ	Gg/m^3	$2.4\ 10^{-3}$
Coefficient of Thermal Expansion	α	$/^\circ C$	$1.0\ 10^{-5}$
Young's Modulus	E_s	MPa	37200.
Poisson's Ratio (Static)	ν_s	-	0.2
Tensile Strength	f_t	MPa	10.
Compressive Strength	f_c	MPa	-36.5
Steel		**Unit**	**Value**
Mass Density	ρ	Gg/m^3	0.
Coefficient of Thermal Expansion	α	$/^\circ C$	0.
Young's Modulus (Static)	E_s	MPa	193000.
Poisson's Ratio (Static)	ν_s	-	0.3
Tensile Strength	f_t	MPa	500.
Compressive Strength	f_c	MPa	-500.
AAR parameters		**Unit**	**Value**
Maximum Volumetric AAR Strain	ϵ^∞	-	28.79×10^{-4}
Characteristic Time	τ_{car}	ATU	3.34
Latency Time	τ_{lat}	ATU	8.29
Activation Energy for Characteristic Time	U_C	K	5,400.
Activation Energy for Latency Time	U_L	K	9,400.
Reference Test Temperature	T_0	$^\circ C$	38.
Tensile Strength	f_t	MPa	10.
Compressive Strength	f_c	MPa	-36.5
Residual Reduction Factor in Tension	Γ_r	-	0.1
Fraction of Tension pre-AAR Comp. Red.	γ_t	-	0.2
Upper Compressive Stress Limit	σ_u	MPa	-10.
Shape Factor for Γ_c	β	-	1.8
Reduction Factor for Young's Modulus	β_E	-	1.
Reduction Factor for Tensile Strength	β_f	-	1.

6. (d) -10 MPa axial, confinement. The initial compressive strain corresponds approximately to the elastic strain (σ/E or 2.7×10^{-4}); then as a result of AAR swelling, it rebounds (especially given that it can only expand axially due to confinement).

7. (b) -10 MPa axial without confinement, -10 MPa axial confinement. The initial compressive strain corresponds approximately to the elastic strain (σ/E or 2.7×10^{-4}); however in the absence of confinement, all of the AAR expansion is redistributed in the radial direction (as opposed to the preceding case).

For their part, the steel radial strains reflect the gradual AAR-induced (swelling) radial strains within the confining jacket.

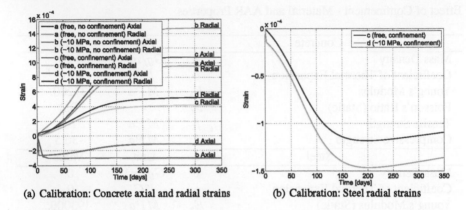

(a) Calibration: Concrete axial and radial strains   (b) Calibration: Steel radial strains

FIGURE 4.13   Calibration: Effect of confinement

Regarding the prediction, Fig. 4.14(a) shows the applied stress history, while Fig. 4.14(b) indicates the corresponding strains. Concrete undergoes an initial elastic axial deformation due to the -5 MPa tension and then expands due to the AAR. When the -10 MPa tension is applied, an elastic strain is produced, at which point the AAR is practically zero since the concrete is being subjected to an axial stress equal to the threshold limit value. When the -10 MPa is lowered to -5 MPa, an elastic "rebounding" once again occurs and the AAR equals zero due to its entire depletion by this time. The concrete radial strain is primarily driven by AAR (a Poisson's effect can be observed but is almost negligible in magnitude). Though partially constrained by the steel jacket, expansion in this case occurs mainly in the radial direction since any axial expansion is being constrained. Moreover, the steel radial strain reflects the time-dependent concrete expansion.

4.1.7   P7: PRESENCE OF REINFORCEMENT

Internal reinforcement inhibits expansion, with AAR-induced cracks then becoming aligned with the reinforcement direction as opposed to traditional "map cracking", (Mohammed et al., 2003). This test problem thus seeks to determine how the numerical model accounts for a restrained (by a reinforcing bar) free expansion, and moreover how internal cracking affects the outcome. No experimental data are available for comparing results. The diameter of the internal steel rebar tested herein is 12 mm.

Concrete is modeled by a nonlinear constitutive model, while a linear elastoplastic model is used for the steel, see Table 4.5.

Fig. 4.15 provides several images of the mesh; let's note that at a distance of approx. 5 mm from the rebar, its effect on AAR is practically nil, and the steel axial stress remains quite small, $\simeq$ 0.063 MPa. Concrete strains (Fig. 4.15) are indeed being restrained in the axial direction, and most of the expansion is occurring in the radial

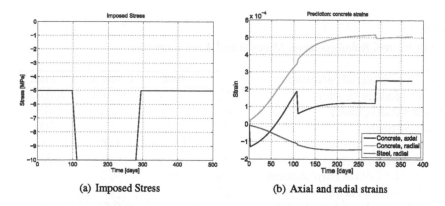

(a) Imposed Stress   (b) Axial and radial strains

FIGURE 4.14   Prediction: Effect of confinement

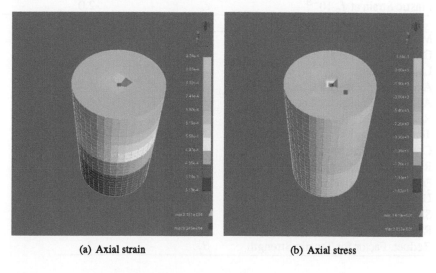

(a) Axial strain   (b) Axial stress

FIGURE 4.15   Effects of reinforcement on AAR

direction. $\varepsilon_{axial} + 2\varepsilon_{radial} = (0.35 + 2(0.85)) \times 10^{-3} = 20.5 \times 10^{-4}$, which is
approximately equal to the specified ε^{∞}.

4.1.8   P8: DAMS

Thus far, the focus has been placed on material response due to individual effects.
In this final case study of Chapter 4, let's consider an idealized dam with a slot. The
dam will be subjected to a series of stressors (i.e. hydrostatic, creep, temperature).

One common remedy for AAR-induced damage in dams consists of cutting a slot
in the structure, as implemented at Mactaquac, (Gilks and Curtis, 2003). This step
serves to relieve the stress state and allow the concrete to expand freely. At some point

TABLE 4.5

Effect of Internal Reinforcement: Material and AAR Properties

Material parameters		Unit	Concrete	Steel
Young's Modulus (Static)	E_s	MPa	37,300.	200,000.
Poisson's Ratio (Static)	ν_s	-	0.2	0.3
Tensile Strength	f_t	MPa	3.5	500.
Fracture Energy 10^{-4}	Γ_f	MN/m	1.00	-
Compressive Strength	f_c	MPa	-38.4	-500.
Compressive Critical Displacement	w_d	m	-0.5	-
Return Direction Factor	β	-	0.	-
Failure Surface Roundness Factor	e	-	0.52	-
Onset of Compression Nonlinearity	f_{c0}	MPa	-13.3	-
Plastic Strain at f_c 10^{-3}	ϵ_c	-	-2.0	-
AAR parameters		Unit	Value	
Max. Vol. AAR Strain $\times 10^{-4}$	ϵ^∞	-	21.9	
Characteristic Time	τ_{car}	ATU	3.34	
Latency Time	τ_{lat}	ATU	8.29	
Activation Energy for Charac. Time	U_C	K	5,400.	
Activation Energy for Latency Time	U_L	K	9,400.	
Reference Test Temperature	T_0	$^\circ C$	38.	
Tensile Strength	f_t	MPa	10.	
Compressive Strength	f_c	MPa	-30.	
Residual Reduction Factor in Tension	Γ_r	-	0.1	
Fraction of Tension pre-AAR Comp. Red.	γ_t	-	0.2	
Upper Compressive Stress Limit	σ_u	MPa	-10.	
Shape Factor for Γ_c	β	-	-2.	
Reduct. Factor for Young's Modulus	β_E	-	1.	
Reduct. Factor for Tensile Strength	β_f	-	1.	

however, concrete swelling will produce contact between the two sides of the slot. This problem therefore will test the model's ability to capture the slot closure and ensuing stress buildup. This last test problem will assess the various types of potential coupling.

The dam is shown in Fig. A.17 with the following conditions: a) the lateral and bottom faces are all fully restrained; b) the front back and top faces are free; c) the slot was cut at time zero with a total thickness of 10 cm; d) the concrete on the right side is reactive, while the concrete block on the left side is not reactive; and e) hydrostatic pressure is applied only on the right block.

Inputting the fitting data from P6, a friction angle of 50° for concrete against concrete and zero cohesion, we consider the two following cases:

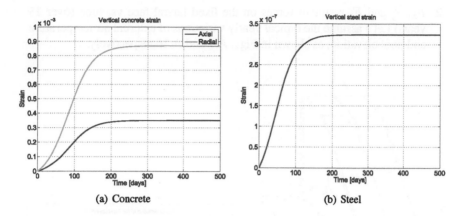

(a) Concrete   (b) Steel

FIGURE 4.16   Effect of Internal Reinforcement

Homogeneous field of internal temperature (20oC), relative humidity (100%) and an empty reservoir

Transient field of external temperature (Fig. A.9), relative external humidity (Fig. A.5) and a pool elevation variation ((Fig. A.16) given by:

$$EL(\text{week}) = \frac{EL_{\max} - EL_{\min}}{2} \sin\left(2\pi \frac{t}{52}\right) + \frac{EL_{\max} - EL_{\min}}{2} \quad (4.6)$$

where $EL_{\max}$ and $EL_{\min}$ equal 95 and 60, respectively.

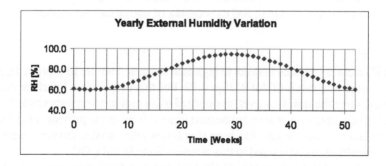

FIGURE 4.17   Humidity variation

For both analyses, the specified temperature and relative humidity are those of the concrete surface, with a zero flux condition between dam and foundation. The reference base temperature of the dam is 20oC.

1. x, y, z displacements of point A

2. F_x, F_y and F_z resultant forces on the fixed lateral face vs. time (over 25
   years) Let's assume the typical yearly variations of external air temperature
   and pool elevation, as shown in Figs. A.9 and A.16 respectively.

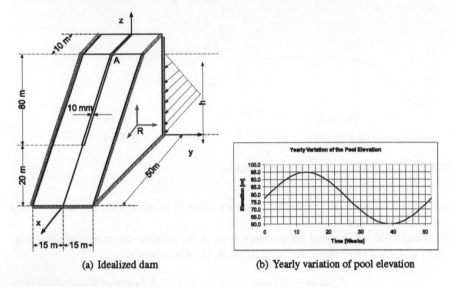

(a) Idealized dam   (b) Yearly variation of pool elevation

FIGURE 4.18   3D dam problem statement

This model seeks to capture: a) the general finite element program capabilities in
modeling the joint response; b) the ease (or difficulty) involved in preparing the input
data file for a realistic problem; and c) coupling of the various parameters.

4.1.8.1 2D

First, a 2D mesh is prepared in order to highlight the code's ability to capture the effect
of slot closure as a result of AAR, Fig. 4.19. For this purpose, an interface element
is inserted along the slot. However, as opposed to an ordinary implementation of this
element, its thickness is not zero but instead finite (10 cm in the present case). Hence,
the formulation of this element has been slightly modified so that the zero stress state
can be considered to correspond to this nonzero finite initial COD.

The mesh is subjected to an (artificially) increasing temperature. In Fig. 4.20, each
line corresponds to a point along the interface. It can thus be observed that contact is
first established at the top. Figs. 4.21(a) and 4.21(b) display the initial mesh with the
10-cm slot and the final mesh after AAR, in highlighting the crack closure process.

The ability of the code to capture slot closure due to AAR is shown in Fig. 4.22(a).
Let's note that by the time of increment 9, the slot has entirely closed (and the corresponding crack opening displacement is indeed 1 cm), along with the presence of a
corresponding stress buildup. By focusing on the first few increments, Fig. 4.22(b), we
can observe the gradual slot closure and corresponding stress increase. In Fig. 4.23(a)

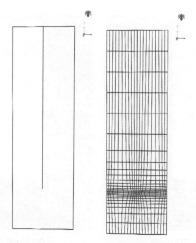

(a) 2D finite element (b) 2D finite element
mesh outline   mesh

FIGURE 4.19   2D Finite element mesh

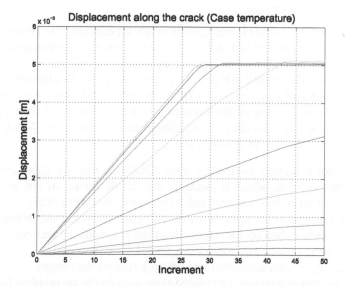

FIGURE 4.20   2D analysis showing slot closure

showing AAR (in its normal evolution), we observe the evolution in slot opening over
time at different elevations. For this case after 12 years, the slot has entirely closed.
Slot displacement is indeed 1 cm since AAR has only been applied on one side of the
dam. Fig. 4.23(b) then shows the corresponding stresses.

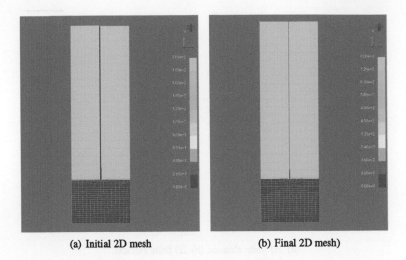

(a) Initial 2D mesh   (b) Final 2D mesh)

FIGURE 4.21   2D slot closure

4.1.8.2   3D case: AAR only

Having successfully captured crack closure in 2D, the question then becomes can this be performed again in 3D? and, if so, what are the potential idiosyncrasies involved in a 3D simulation that cannot be captured by a mere 2D simulation?

The 3D mesh of the model in Fig. A.17 is shown in Fig. 4.24. In this initial analysis, the dam has been subjected only to AAR, as shown in Table 4.6; for the 10-mm slot with AAR, we are seeking to determine slot thickness in terms of time (as reduced by AAR), Fig. 4.25(a), and the corresponding contact stresses, Fig. 4.25(b).

From these plots, we can conclude that:

1. The slot is initially completely open (COD=0); then the COD gradually reaches -10 mm, which is precisely the slot thickness. At this point, the interface element is activated and no more expansion can occur.
2. The concavity of the surfaces is to be recognized. Greater expansion occurs at the center than on the edges. Eventually, expansion becomes entirely uniform and nearly full contact is achieved at -10 mm.
3. Similarly, less expansion occurs at the top than on the bottom.
4. The bottom also undergoes reduced expansion due to the problem formulation, since it is being constrained by the bottom concrete.
5. Similarly, stresses equal zero at the outset due to a lack of contact.
6. The stress gradually increases and we observe the same concavity as that noted for the COD. In other words, stresses are much greater at the center than on the edges.
7. Concavity remains present even after many years, with stresses being greater in the central part than on the edges.

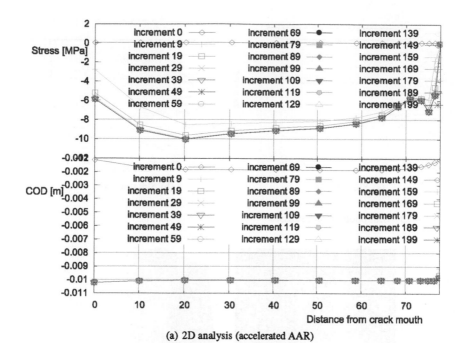

(a) 2D analysis (accelerated AAR)

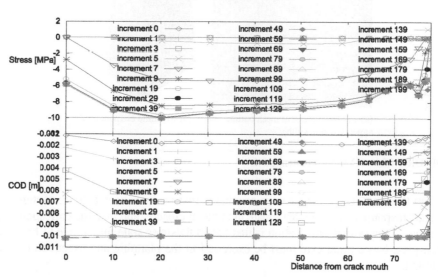

(b) 2D analysis (accelerated AAR)

FIGURE 4.22   2D analysis: Crack closure

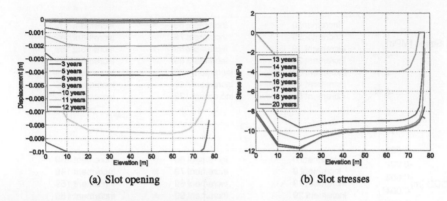

(a) Slot opening   (b) Slot stresses

FIGURE 4.23   Slot opening and stresses

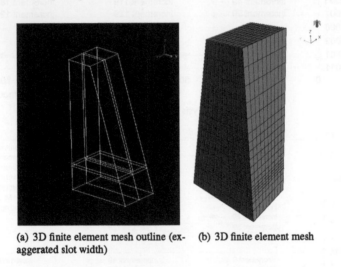

(a) 3D finite element mesh outline (ex-   (b) 3D finite element mesh
aggerated slot width)

FIGURE 4.24   3D Finite element mesh

From these observations, it can be concluded that we have been able to: capture the
"true" response of the slot, anticipate the time of closure, and map the corresponding
stresses. One salient observation is that the COD/stress state may prove to be quite
complex, hence great care must be exercised in the planning and execution of slot
cutting.

4.2 SUMMARY

The engineering community has fallaciously pursued unrealistic benchmark problems
which are too complex and make it impossible to point specific deficiencies of a model.
Instead, one should validate an AAR model with a battery of small problem, each
assessing the code capability to capture a specific idiosyncracy of AAR.

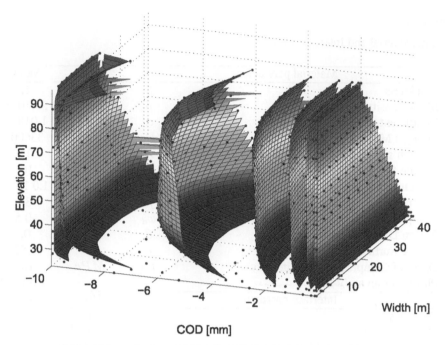

(a) Slot thickness for times 0.167, 1.67, 3.33, 5, 6.67, 8.33, 10 and 11.67 years

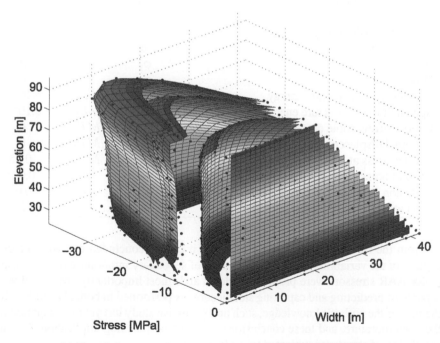

(b) Normal slot stresses for times 8.33, 11.67, 15, 18.33 and 121.67 years

FIGURE 4.25   3D response of a dam subjected to AAR

TABLE 4.6

Dam: Material and Interface Properties

Material parameters		Unit	Value
Thickness	h	m	1.
Mass Density	ρ	Gg/m^3	2.410^{-3}
Coefficient of Thermal Expansion	α	$/^{\circ}C$	1.010^{-5}
Young's Modulus (Static)	E_s	MPa	33600.
Poisson's Ratio (Static)	ν_s	-	0.2
Tensile Strength	f_t	MPa	10.
Fracture Energy	Γ_f	MN/m	$1.0\ 10^{-4}$
Compressive Strength	f_c	MPa	-30.
Compressive Critical Displacement	w_d	m	-0.5
Return Direction Factor	β	-	0.
Failure Surface Roundness Factor	e	-	0.52
Onset of Compression Nonlinearity	f_{c0}	MPa	-13.3
Plastic Strain at f_c	ϵ_c	-	$-2.0\ 10^{-3}$
Interface parameters		Unit	Value
Thickness	h	m	1.
Mass Density	ρ	Gg/m^3	0.
Coefficient of Thermal Expansion	α	$/^{\circ}C$	0.
Tangential stiffness	K_t	MPa	1.
Normal stiffness	K_n	MPa	1.10^8
Uniaxial tensile Strength	σ_t	MPa	0.1
Cohesion	c	-	0.1
Angle of friction	ϕ_f	$^{\circ}$	10.
Dilatancy angle	ϕ_D	$^{\circ}$	0.
Specific mode I fracture energy	G_F^I	MJ	5.10^{-03}
Specific mode II fracture energy	G_F^{IIa}	MJ	5.10^{-03}
Ratio of irreversible deformation	γ	MPa	0.
Maximal displacement for dilatancy	u_{Dmax}	MPa	0.
Tensile stress at break-point for the bilinear law	σ_1	MPa	0.05
COD at break-point for the bilinear law	sw_1	m	0.025
Cohesion at break-point for the bilinear law	c_1	-	0.05
CSD at break-point for the bilinear law	cw_1	m	0.025

From the results presented in this chapter, it can be concluded that with the exception of the variability in relative humidity (of secondary importance for dams), all major AAR stressors were properly modeled and, most importantly, the model was capable of predicting and capturing slot closure, as performed in both 2D and 3D. To the best of the author's knowledge, such an exhaustive study has yet to be reported in the open literature, and these conclusions can have far-reaching implications for the simulation of structures subjected to AAR.

5 Parametric Study

The constitutive model presented in Chapter 2 has been "validated" in Chapter 4, yet one question remains unanswered: how sensitive is the numerical model to the various modeling assumptions that can now be introduced? Though this chapter will focus on dams, most of its conclusions can be extended to other types of structures (e.g. a nuclear reactor) as well, albeit in accounting for the appropriate engineering-based adjustments.

The questions targeted in this chapter will include:

1. How does AAR affect the stress state in a dam subjected to normal gravity, hydrostatic and temperature loads (H, G, and T respectively)?
2. Should we account for the seasonal variation in pool elevation and temperature during a long-term analysis of a dam subjected to AAR?
3. How many increments per year are necessary to obtain reliable results?
4. Is it important to model the dam/foundation joint in an AAR analysis?
5. How important is the modeling of sigmoid kinetics?
6. How significant is the degradation effect on the Young's modulus value of concrete?
7. Should both the internal and external concretes of a dam be modeled?
8. How different is this model from simpler versions?

5.1 PRELIMINARY

5.1.1 PROBLEM DEFINITION

An actual arch-gravity dam was selected for this parametric study (the same dam will be subsequently analyzed as a case study in Chapter 7 and has been illustrated in Fig. 5.1). Only a 2D analysis will be conducted herein. Table 5.1 summarizes the 19 analyses undertaken.

5.1.2 PRIMARY UNITS

Great care must be exercised to ensure use of consistent units in nonlinear analysis, and even more so for time-dependent analyses (e.g. when AAR is involved). In the following parametric study, the units listed in Table 5.2 will be adopted. One particularly challenging task consists of handling the time variable since each increment must correspond to a unit time. This challenge has been met by introducing a so-called *Analysis Time Unit* (ATU), which is set equal to (unless otherwise noted): 2 units/month (2.17 weeks or 15 days). Each ATU will feature a variation of:

- Material properties: AAR expansion and material degradation (Young's modulus and tensile strength)

TABLE 5.1

Summary of parametric studies

Load	AAR Model	R/C Joint	
		No	Yes
G+T		√	
G+T+H		√	
	S-P	√	√
	S-P, ER	√	
	SC, SE	√	√
G+T+H+A	SC, LE	√	√
	S-P, 12 incr./year	√	
	S-P, 4 incr./year	√	
	S-P, 2 concrete types	√	
G+T+H'+A	S-P	√	√
G+T+A	S-P	√	√
G+T'(7º)+A	S-P	√	
G+T'(4º)+A	S-P	√	
G+T'(10º)+A	S-P	√	

G	Body force
T	Temperature variable
T'	Temperature constant
H	Hydrostatic variable
H'	Hydrostatic constant
A	AAR load.
S-P	Saouma/Perotti's model
SC	A simplistic model
ER	Elastic modulus reduction
SE	Sigmoid expansion
LE	Linear expansion

TABLE 5.2

Units adopted in this parametric study

Force	MN
Length	m
Time	s
Pressure	MPa
Mass density	Gg/m^3
Weight density	$Gg/(m^2 s^2)$; or MN/m^3
Fracture energy	MJ/m^2 (MN/m)

- Load: Pool elevation, uplift pressures, crack lengths, air/water (and hence concrete) temperature

5.1.3 ELASTIC AND THERMAL PROPERTIES

As with most arch dams, the one considered herein uses two concrete mixes, one for the surface the other for the inner core. Surface concrete has a higher cement content, resulting in greater strength (and elastic modulus) and thus greater potential for AAR expansion. The default elastic and thermal properties adopted in this study are shown

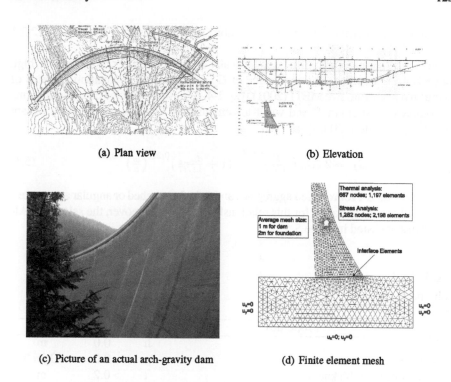

(a) Plan view (b) Elevation

(c) Picture of an actual arch-gravity dam (d) Finite element mesh

FIGURE 5.1 Dam used for this parametric study

in Table 5.3. Let's note that the concrete, as well as the rock, is assumed to be linear elastic, with the only non-linearity stemming from the presence of an interface joint between concrete and rock.

TABLE 5.3

Concrete elastic and thermal properties

Material properties		Units	Concrete			Rock
			Ext.	Int.	Average	
Cement content		Kg/m³	250	190	-	
Thickness	h	m	1.0	1.0	1.0	1.0
Mass density	ρ	Gg/m³	2.50E-3	2.50E-3	2.50E-3	0.0
Coeff. of thermal exp.	α	/oC	9.50E-6	9.50E-6	9.50E-6	-
Young's modulus	E_s	MPa	31,670	29,380	30,525	24,000
Poisson's ratio	ν_s		0.167	0.167	0.167	0.250
Compressive strength	f_c	MPa	-35.0	-26.0	-30.5	-
Tensile strength	f_t	MPa	2.0	1.6	1.8	-
Diffusivity	h	m²/s	1.45E-06			
Conductivity	k	MN.m/(m.ATU.°C)	3.87			-
Specific heat	c	MN.m/(Gg.°C)	887			-
Density	ρ	Gg/m³	2.50E-3			-
Base temperature	T	°C	7			-

The joint properties are displayed in Table 5.4 and equations from Sect. 3.2.2.1 were used.

Since the normal joint stiffness is basically a penalty number, a reasonably high value is to be assumed. The expression of G_I^F is an empirical equation in terms of maximum aggregate size METI (2001) (Eq. 2-4-13). An alternative equation in terms of compressive strength f_c' and water-to-cement ratio w/c was proposed by Bažant and Becq-Giraudon (2001):

$$G_F = 2.5\alpha_0 \left(\frac{f_c'}{0.051}\right)^{0.46} \left(1 + \frac{d_{max}}{11.27}\right)^{0.22} \left(\frac{w}{c}\right)^{-0.30} \tag{5.1}$$

In this case, α_0 is 1 for rounded aggregates and 1.44 for crushed or angular aggregates. The coefficient of variation for this relationship is 30%. Moreover, the selected AAR properties are listed in Table 5.5.

TABLE 5.4
Joint properties

Properties		Values	Units
Thickness	h	0.0	m
Mass density	ρ	0.0	Gg/m3
Cross interface thickness	t	0.2	m
Coefficient of thermal expansion	α	0.0	
Static Young's modulus	E	2.40E+04	MPa
Tangential stiffness	K_t	1.20E+05	MPa
Normal stiffness	K_n	1.20E+05	MPa
Compressive strength	f_c	30.5	MPa
Tensile strength	f_t	1.806	MPa
Cohesion	c	1.500	MPa
Friction angle	ϕ	35.0	degree
Dilatancy angle	FDs	20.0	degree
Maximum aggregate size	d_{max}	0.08	m
Fracture energy mode I	G_I^F	3.13E-04	MN/m
Fracture energy mode II	G_{II}^F	3.13E-03	MN/m
Relative value of irreversible deformation	γ	0.3	
Maximal displacement for dilatancy	u_{max}^D	0.01	m
Tensile stress at the break-point	s_1	0.452	MPa
Crack opening displacement at the break- point	s_{w1}	1.73E-04	m
Cohesion at the break-point	c_1	0.375	MPa
Crack sliding displacement at the break-point	c_{w1}	1.56E-03	m

TABLE 5.5

AAR properties for this parametric study

Property	External	Internal	Average	
Thermodynamic properties				
Char. time	2.7	1.4	2.1	ATU
Lat. time	4.9	4.3	4.6	ATU
Vol. exp.	5.4E-03	4.5E-03	4.9E-03	-
Activation energy for characteristic time		5,400		°K
Activation energy for latency time		9,400		°K
Reference test temperature		60		°C
Strength				
Fictitious tensile strength		10.00		MPa
Compressive strength		-30.50		MPa
Γ				
Residual reduction factor in tension		0.10		-
Fraction of tension pre-AAR reduction		0.18		-
Upper compressive stress limit		-10.00		MPa
Shape factor for Gamma_c		0.50		-
Degradation				
Reduction factor for Young's modulus		1.0		-
Reduction factor for tensile strength		1.0		-

5.1.4   PRELIMINARY THERMAL ANALYSIS

Given that our test problem is an arch dam at high altitude, a preliminary (linear) thermal study needs to be performed. This analysis period spans 10 years, in extracting the spatial and temporal temperature distribution for the last year (after verification of little variation with respect to the previous year). Such an analysis is warranted so as to account both for the effect of temperature in the AAR expansion (through Equation 2.5) and for thermal stresses. Be advised however that regarding the temperature effect, the total temperature difference (as converted to °K) is necessary, whereas the assessment of thermal stress effect requires ΔT, i.e. the actual temperature minus the reference temperature ($7^\circ C$).

The external temperatures are shown in Fig. 5.2(a). Let's note that the rock has been set at $3^\circ C$; moreover, the five temperature distributions for the downstream face are due to variations in water temperature with depth (Eq. 6.10), while just a single distribution exists for the upstream face (which, unsurprisingly, is much warmer than the downstream face).

Results are then compared with *in situ* temperature sensors at selected points, see Fig. 5.2(e); a strong correlation can be observed for two of the three sensors.

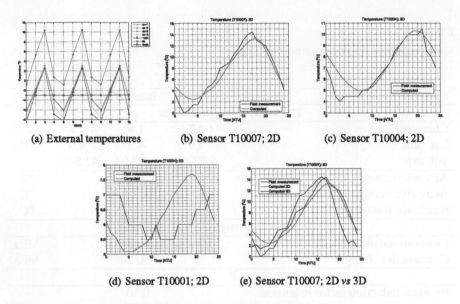

(a) External temperatures   (b) Sensor T10007; 2D   (c) Sensor T10004; 2D

(d) Sensor T10001; 2D   (e) Sensor T10007; 2D *vs* 3D

FIGURE 5.2   Computed temperature *vs.* field measurements

Furthermore, results of a 2D analysis contrast with findings from a 3D analysis, though no conclusive evidence is available to suggest one is better than the other.

The final temperature distribution for the central section is shown in Figure 5.3 for each of the twelve representative months of the year.

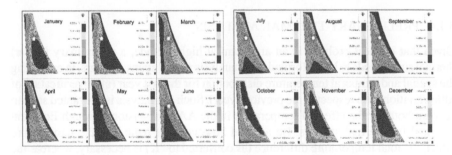

FIGURE 5.3   Yearly temperature variation (7°C should be added)

5.2   RESULTS

Results of the various parametric analyses will be presented next. The loadings will be labeled G, T, H, and A for gravity, thermal, hydrostatic and AAR respectively.

5.2.1 WITHOUT A FOUNDATION/DAM INTERFACE

Let's begin by considering the simplified 2D case in which no joint elements are inserted between the rock and the concrete.

5.2.1.1 (G+T+H)-(G+T); Role of the hydrostatic load

The role of the hydrostatic load on absolute and relative crest displacements is depicted in Figures 5.4(a) and 5.4(b). A 50 When the minimum and maximum principal stresses

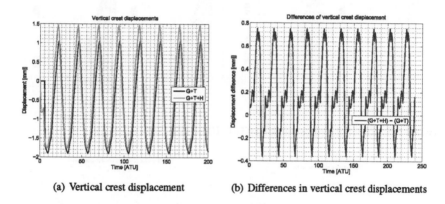

(a) Vertical crest displacement   (b) Differences in vertical crest displacements

FIGURE 5.4   Effect of hydrostatic load (G+T+H)-(G+T) Δ v

are compared (Figs. 5.5(a) and 5.5(b) respectively), the hydrostatic load is seen to exert only a small influence on the minimum, while the maximum tensile stress is reduced. Though this last observation may be counterintuitive, let's still recall that the thermal load is being included.

5.2.1.2 (G+T+H+A)-(G+T+H); Role of AAR expansion

The role of AAR expansion is displayed in Figures 5.6(a) and 5.6(b). In this study, AAR expansion is initiated at about 300 ATU (approx. 12 years), with the expansion being shown here for 1,200 ATU (i.e. 50 years): as expected, AAR causes an irreversibly increasing crest displacement. The effects of seasonal variations are also presented. Figures 5.7(a) and 5.7(b) indicate the maximum and minimum principal stresses, respectively. AAR serves to increase both of them; however, it should be noted that due to AAR expansion, the core of the dam is subjected to tensile stresses (this finding will be revisited later once a crack has been allowed to develop between the dam and the rock). This internal tensile stress will occur in practically all subsequent analyses and may cause internal cracking (which may be noticeable from the gallery). Attention should be paid to the possible "daylighting" of these cracks on the downstream side of the dam.

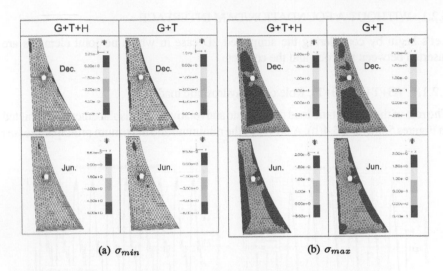

(a) σ_{min} (b) σ_{max}

FIGURE 5.5   Effect of hydrostatic load (G+T+H)-(G+T); $\Sigma M_{min}, \sigma_{max}$

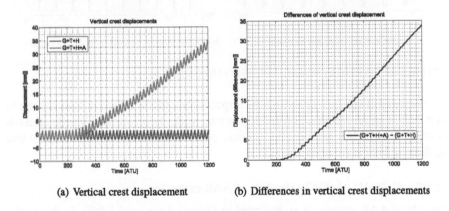

(a) Vertical crest displacement   (b) Differences in vertical crest displacements

FIGURE 5.6   Effect of AAR (G+T+H+A)-(G+T+H) Δ v

5.2.1.3   (G+T+A+H)-(G+T+A): Role of the hydrostatic load (revisited)

The role of the hydrostatic load, in conjunction with both thermal and AAR expansion, will be examined next. As opposed to Figure 5.4(a), Figure 5.8 reveals the small influence of hydrostatic load on crest displacement. This difference increases over time as a result of the coupling between AAR and stress. Once again, the minimum and maximum stresses are barely affected by the presence of hydrostatic load and AAR expansion, see Fig. 5.9.

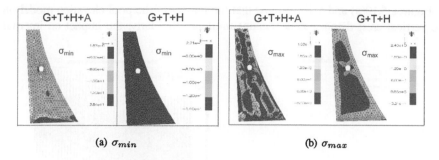

(a) σ_{min} (b) σ_{max}

FIGURE 5.7   Effect of hydrostatic load (G+T+H+A)-(G+T+H); ΣM_{min}, σ_{max}

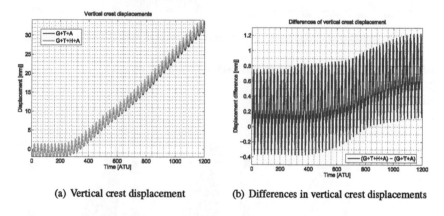

(a) Vertical crest displacement   (b) Differences in vertical crest displacements

FIGURE 5.8   Effect of hydrostatic load (G+T+H+A)-(G+T+A); Δ v

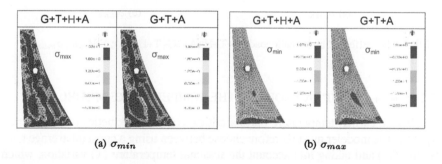

(a) σ_{min} (b) σ_{max}

FIGURE 5.9   Effect of hydrostatic load (G+T+H+A)-(G+T+A); ΣM_{min}, σ_{max}

5.2.1.4   (G+T+H+A)-(G+T+H'+A): Role of the hydrostatic model

The effect of pool elevation variation will now be addressed. In an exact analysis (H), the pool elevation is seasonally adjusted, whereas according to a simplified model (H'), pool elevation is set at a constant value equal to the average elevation.

Figures 5.10(a) and 5.10(a) show that the impact of the pool elevation model on the vertical crest displacement remains minimal. Similarly, the impact on internal

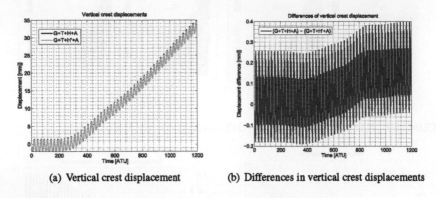

(a) Vertical crest displacement   (b) Differences in vertical crest displacements

FIGURE 5.10   Effect of pool elevation (G+T+H+A)-(G+T+H'+A); Δ v

stresses is also minimal.

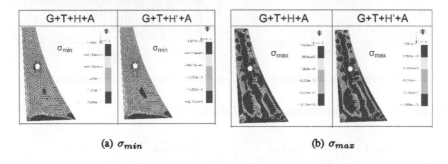

(a) σ_{min} (b) σ_{max}

FIGURE 5.11   Effect of hydrostatic load (G+T+H+A)-(G+T+H'+A); $\Sigma M_{min}, \sigma_{max}$

5.2.2   (G+T+A)-(G+T'+A): ROLE OF THE TEMPERATURE MODEL

As indicated earlier, temperature plays a critical role in the kinetics of the reaction (Eq. 2.5). The modeler must therefore choose between using a constant average temperature (T') and taking into account the seasonal temperature (T) variation, which is presented in Figure 5.2(a).

The temperature effect is immediately apparent from Figure fig:param-11. As expected, the higher the constant temperature, the faster the reaction; more importantly however, the variable temperature slows the expansion to a lower peak. This last observation was indeed unanticipated (i.e. a constant temperature yielding higher vertical crest expansion) and may be attributed both to the temporal evolution in compressive stresses (which inhibit expansion) and to the reduction in AAR volumetric expansion due to the presence of tensile stresses (which are more likely to occur with T).

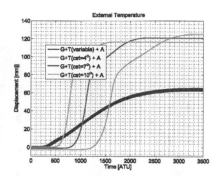

FIGURE 5.12   Effect of temperature variation (G+T+A)-(G+T'+A); Δv

Stress contour plots at 50 years are shown in Figure 5.13(a). In recognition of the fact that these plots do not correspond to identical vertical crest displacements, the constant temperature (T') errs away from the conservative side (i.e. in favor of lower minimum and maximum stresses).

In comparing the two analyses for the same displacement and time increment (i.e. 43 years or 1,032 ATU), where T' for $7oC$ intersects T (variable temperature), the discrepancy between the resulting stresses of these two analyses becomes even more accentuated: T' causes an even larger decrease in compressive and tensile stresses, see Fig. 5.13(b). This result is corroborated by Figure 5.13, which compares T and T' ($7oC$) at 65 years (1,560 ATU).

5.2.2.1   (E)-(E'): Effect of concrete deterioration

AAR causes concrete deterioration through a gradual decrease in both Young's modulus E and tensile strength f'_t, see Fig. 2.15.

In assuming a 30% deterioration, the vertical and horizontal crest displacements are shown in Figure 5.14.

Let's note that while concrete deterioration exerts a relatively small impact on vertical displacements, such is not the case for horizontal displacements.

Moreover, it should be kept in mind that these are 2D plane strain analyses and AAR expansion will induce high compressive stresses in the lateral direction; consequently, a portion of the vertical crest displacement is due to the Poisson's effect. In conclusion and as expected, concrete deterioration will reduce both the tensile and compressive stresses, Fig. 5.15.

5.2.2.2   (G+T+H+A)-(G+T+H+A'): Effect of modeling internal and external concretes

Arch dams feature a thin layer of high cement content concrete on both the upstream and downstream faces. High cement content undoubtedly leads to increased AAR expansion according to the properties listed in Table 5.5.

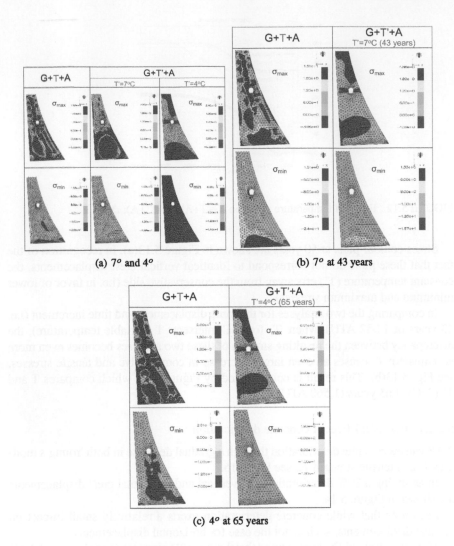

(a) 7° and 4°

(b) 7° at 43 years

(c) 4° at 65 years

FIGURE 5.13   Effect of temperature variation (G+T+A)-(G+T'+A); σ_{max} and σ_{min}

Figure 5.16 clearly shows that the difference in vertical expansion is negligible (and, as expected, modeling the thin layer of high cement content concrete results in a slightly higher expansion). Similarly, the stresses are nearly identical, with the exception of the external concrete (upstream and downstream).

5.2.2.3   (G+T+A): Effect of time discretization

At present, most of the critical model parameters have been identified. Using these parameters (or modeling assumptions), the next step entails examining the effect of

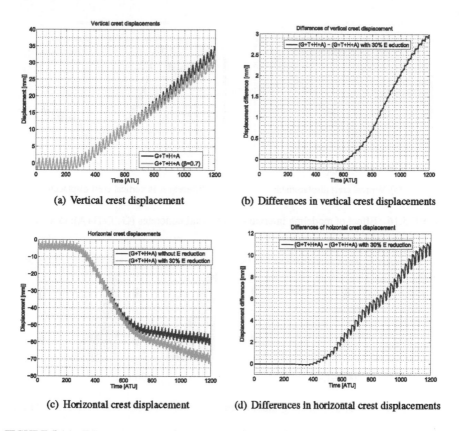

(a) Vertical crest displacement

(b) Differences in vertical crest displacements

(c) Horizontal crest displacement

(d) Differences in horizontal crest displacements

FIGURE 5.14 Effect of concrete deterioration; Δ v and Δ h

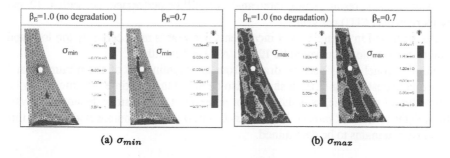

(a) σ_{min}

(b) σ_{max}

FIGURE 5.15 Effect of concrete deterioration (G+T+H+A)-(G+T+H'+A); $\Sigma M_{min}, \sigma_{max}$

time discretization Δt. Since AAR simulation is highly nonlinear required to span at least 20 years, it is of paramount importance to determine the large time step capable of yielding a reasonable result. As opposed to an explicit dynamic time integration, it is practically impossible to determine, in such an analysis, the maximum time step

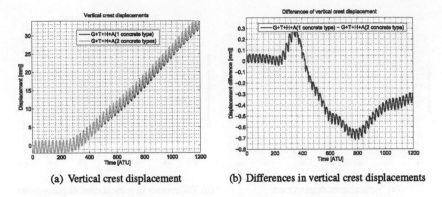

(a) Vertical crest displacement   (b) Differences in vertical crest displacements

FIGURE 5.16   Effect of modeling internal and external concretes (G+T+H+A); Δ v

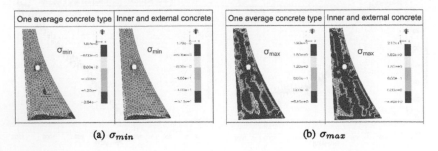

(a) σ_{min} (b) σ_{max}

FIGURE 5.17   Effect of modeling internal and external concretes (G+T+H+A); ΣM_{min}, σ_{max}

to ensure a stable result. Hence, three analyses will be performed, i.e. with 4, 12 and 24 increments (ATU) per year.

As displayed in Figure 5.18, 4 increments per year is most certainly too low, and a more reasonable value would be 12 (i.e. ATU = 1 month).

Similar conclusions may be drawn from an examination of the minimum and maximum stresses, Fig. 5.19. It should also be noted that the time increment may be a function of mesh size, as is the case for an explicit time history dynamic analysis through the so-called *Courant–Friedrichs–Lewy condition* (Courant et al., 1967); this conclusion remains to be determined.

5.2.2.4   Role of the kinetic model

The type of expansion model adopted is most critical to this study. For many years, the state of knowledge and state-of-the-art were both encapsulated in the model first introduced by Hayward et al. (1988) (based on the work of Grob (1972) and Wittke and M. (2005)), Fig. 2.7 (refer to Section 2.4.1) and later adopted by Charlwood et al. (1992), Thompson et al. (1994) and Curtis (1995)). This model does not explicitly

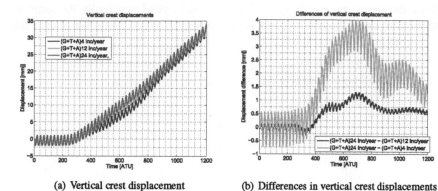

(a) Vertical crest displacement   (b) Differences in vertical crest displacements

FIGURE 5.18   Effect of time discretization (G+T+A); Δ v

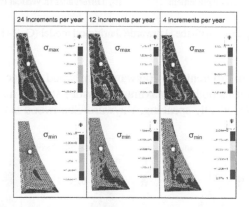

FIGURE 5.19   Effect of time discretization (G+T+A); $\Sigma M_{min}, \sigma_{max}$

address reaction kinetics (i.e. time dependence), but simply considers the uniaxial expansion reduction due to normal compressive stress in the concrete according to:

$$\dot{\epsilon} = \dot{\epsilon}_{go}(t) - K \log(\sigma_i/\sigma_0) \qquad (5.2)$$

where $\dot{\epsilon}_{go}(t)$ is the unrestrained time-dependent concrete growth rate at low stress, K the slope of the line defining the concrete growth rate vs. log stress, σ_i the principal (i=1-3) stresses corresponding to the three principal stress directions, and σ_0 a low stress level above which the concrete growth strain is reduced.

Given this model's continued prominence (mostly by engineering firms), we will examine the differences in results obtained between this simplified model and the author's model presented in Chapter 2 (identified herein as Saouma/Perotti).

In order to compare these two models, the Hayward/Charlwood model was assigned the same kinetics as the Saouma/Perotti model, though none of the other features from the model were replicated. Based on Figure 5.20, it would be tempting to draw the surprising conclusion that both models yield practically the same results. In comparing the resulting stress field however, Fig. 5.21(a), we remark that

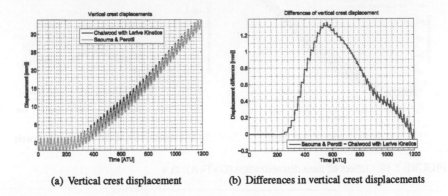

(a) Vertical crest displacement   (b) Differences in vertical crest displacements

FIGURE 5.20   Comparison with the Hayward/Charlwood model (G+T+H+A), same kinetics; Δv

they are substantially different: the Hayward/Charlwood is much less conservative in that it underestimates both the maximum (tensile) stresses and minimum (compressive) stresses. Practically speaking, Hayward/Charlwood is unlikely to be used

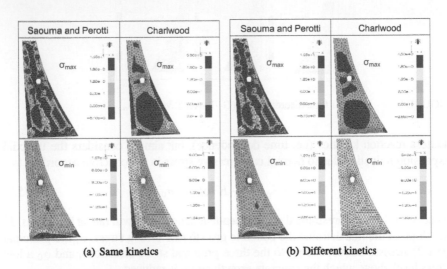

(a) Same kinetics   (b) Different kinetics

FIGURE 5.21   Comparison with the Hayward/Charlwood model (G+T+H+A), same kinetics; ΣM_{min}, σ_{max}

with Larive kinetics. Consequently, the procedure adopted is similar to the one often followed in practice, whereby final expansions are calibrated to yield identical vertical displacements after 50 years. The Saouma/Perotti model uses its (Larive-based) sigmoid kinetics, whereas Hayward/Charlwood employs a linear form. Figure 5.22 shows identical final expansions (as expected), yet a substantial difference exists in the time-dependent expansion. As regards the resulting stresses, Figure 5.21(b) once again reveals that the Hayward/Charlwood model is still much less conservative; moreover, the discrepancy with the author's more refined model is even greater than when both models have assumed the same kinetics.

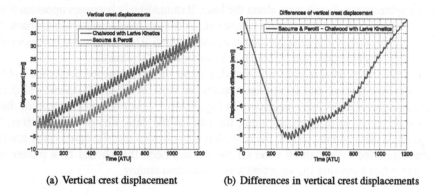

(a) Vertical crest displacement   (b) Differences in vertical crest displacements

FIGURE 5.22   Comparison with the Hayward/Charlwood model (G+T+H+A), different kinetics

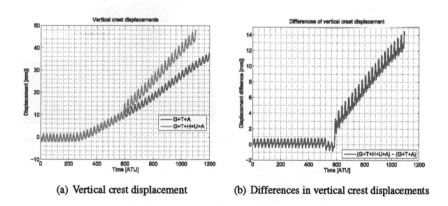

(a) Vertical crest displacement   (b) Differences in vertical crest displacements

FIGURE 5.23   Effect of pool elevation (G+T+H+U+A)-(G+T+A); Δ v

5.2.3   MODEL WITH INCLUSION OF JOINT

In the preceding section, all analyses considered a rigid rock/concrete joint. In reality, such is never the case as this interface is indeed likely to crack and possibly even slide.

In utilizing the preliminary conclusions from previous runs, this section will examine the effect of various parameters on a model that accommodates potential crack-
ing/sliding through the use of interface elements, (Slowik et al., 1998; Červenka et al.,
1998).

Let's begin by repeating the analysis of the dam subjected to G+T+H+A and
considering two cases without joints (Figs. 5.24(a) and 5.24(b)) and with joints (Fig.
5.24(c)). In the absence of an interface, the best course of action is to plot the stress
(and displacement) distribution along the base. It immediately becomes apparent that
the interface is subjected to tensile stresses at the center; for all practical purposes, it
is as if the dam were simply supported at its toe and heel. This finding can be observed
from the v displacement and principal contour lines of the stresses.

If on the other hand a joint is inserted, we can then plot for the joints the uplift
pressures, the normal and tangential stresses, as well as the crack opening/sliding
displacements (COD/CSD), as shown in Fig. 5.24(c). It is even more apparent that
a zone of tensile stresses exists along the core of the dam, along with a "lift-off"
between the dam and the rock at the base. It is also obvious that the dam is only really
being supported over a small zone on the toe and heel; this should be a matter of
concern when assessing the structure's safety against sliding.

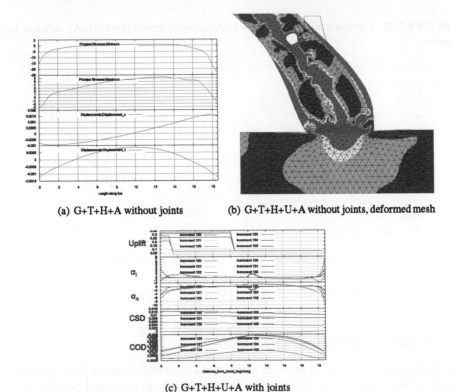

(a) G+T+H+A without joints   (b) G+T+H+U+A without joints, deformed mesh

(c) G+T+H+U+A with joints

FIGURE 5.24 Effect of joints

5.2.3.1 Effect of hydrostatic load

Having determined the importance of joints, let's now revisit the effect of hydrostatic load (previously examined in Section 5.2.1.4. From Figure 5.25, we once again note an identical set of results for 25 years (600 ATU). At this point, it is clear that sudden sliding has occurred; through dilatancy of the interface joint, an accompanying sudden increase in the vertical displacement is also observed. In contrast, only little

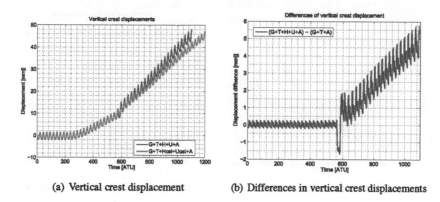

(a) Vertical crest displacement   (b) Differences in vertical crest displacements

FIGURE 5.25   Effect of hydrostatic load (G+T+H+U+A)-(G+T+H'+U'+A) on Δ v

influence seems to be exerted by the hydrostatic load on both the internal maximum and minimum stresses, Fig. 5.26.

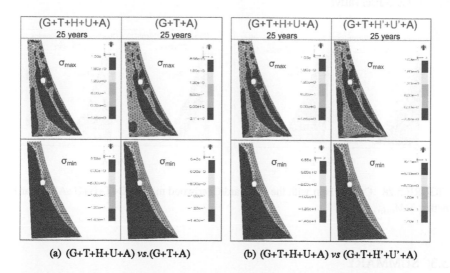

(a) (G+T+H+U+A) vs.(G+T+A)   (b) (G+T+H+U+A) vs (G+T+H'+U'+A)

FIGURE 5.26   Effect of hydrostatic load (G+T+H+U+A)-(G+T+H'+U'+A), $\Sigma M_{min}, \sigma_{max}$

5.2.3.2 Effect of the kinetic model

We will next revisit a jointed model with (G+T+H+U+A), yet with a different kinetic model (as previously examined in Section 5.2.2.4 for the non-jointed model). Two kinetic models have been considered: the author's, and a linear model for Hayward/Charlwood that has been calibrated to yield the same vertical displacement at year 25 (as shown in Fig. 5.27). As identified above, not only are the vertical crest

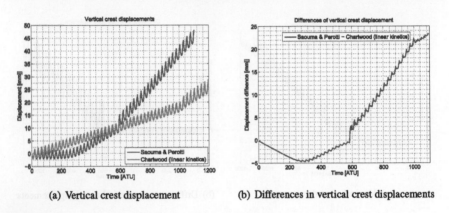

(a) Vertical crest displacement   (b) Differences in vertical crest displacements

FIGURE 5.27 Comparison with the Hayward/Charlwood model (G+T+H+U+A), linear kinetics; Δv

displacements quite different, but the maximum principal stresses, Fig. 5.28, remain much less conservative.

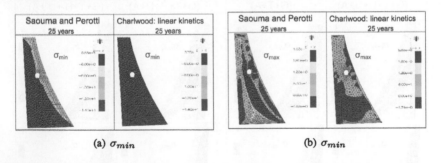

(a) σ_{min} (b) σ_{min}

FIGURE 5.28 Comparison with the Hayward/Charlwood model (G+T+H+U+A), linear kinetics; $\Sigma M_{min}, \sigma_{max}$

5.3 SUMMARY

Following this extensive parametric study, a set of preliminary recommendations can be drawn for the numerical modeling of dams subjected to AAR, namely:

1. The model must contain a temporal and spatial variation of temperature $(T(x, y, z, t)$; use of a constant average temperature must be avoided.
2. A hydrostatic/uplift load is important only in assessing failure, as the effect on AAR-induced responses is minimal.
3. 12 or 24 increments per year are required in order to correctly capture the effects of temperature variation.
4. The internal and external concretes do not need to be modeled separately.
5. The so-called "log model" of Hayward et al. (1988), when calibrated to yield the same displacements (and linear kinetics) as Saouma/Perotti, outputs stresses that are much less conservative.
6. AAR causes a bulge in the displacement and tension at the center. Such phenomena are critical since they may cause undetectable internal cracks and explain the cracks observed inside some dams.
7. The foundation/dam interface must also be modeled.

6 Material Properties

The selection of material properties is of paramount importance. According to the old adage *garbage in, garbage out*; in other words, nonsensical input data will result in equally nonsensical output (all too often, unfortunately, engineers fail to perform a "sanity" check of the results to determine if they are indeed reasonable).

This chapter will provide guidance on the selection of material properties for a nonlinear AAR study and will rely extensively on a study conducted by the US Bureau of Reclamation on existing dams suffering from AAR.

6.1 INTRODUCTION

Material properties can be divided into two sets, those pertaining to the constitutive model (stress-strain, whether linear or nonlinear) and crack fracture energy; and those pertaining to the AAR model. Since the proposed AAR model (see Chapter 2) can accommodate any type of nonlinear stress-strain, let's limit the discussion herein to the most fundamental (and critical) parameters: modulus of elasticity E, tensile strength f'_t and compressive strength f'_c on the one hand, and τ_l, τ_c and $\varepsilon(\infty)$ on the other. These choices will be covered in Sections 6.2 and 6.3, respectively.

As shown in Chapter 5, a thermal analysis must be conducted in order to determine the spatial and temporal distributions of temperature; hence, concrete thermal properties will also be addressed in Section 6.4.

Section 2.6.3 displayed that AAR reduces the modulus of elasticity E and tensile strength f'_t, both of which are critical material input parameters for a numerical simulation. It remains exceedingly difficult however to extrapolate laboratory-based deterioration parameters to field values. Section 6.5 will thus be devoted to an extensive study by the Bureau of Reclamation, which discusses the actual, field-measured deterioration of concrete properties caused by AAR. This discussion may provide guidance on how to assign values to β_E and β_f in Eqs. 2.28 and 2.29.

Section 6.6 will then present a numerical algorithm for the System Identification of AAR properties based on field observations of crest displacements or other measurable quantities.

6.1.1 ON THE RANDOMNESS OF PROPERTIES

The selection of material properties should take into account their inherent spatial variability within a structure (where concrete, possibly from various sources, is likely to have been cast over an extended period of time with less than perfect quality control).

A variation in test results is expected due to multiple factors, including the method for placing and aging concrete under field conditions. As a case in point, Figure 6.1 illustrates typical compressive stress-strain curves, as measured on 15 core samples taken from a single Reclamation dam and then tested at Reclamation laboratories according to ASTM standards, (Mills-Bria et al., 2006). In another related study, the

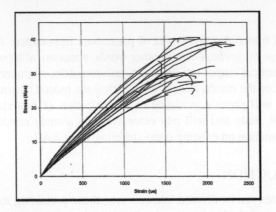

FIGURE 6.1   Variation in 15 compressive stress-strain tests conducted at the same dam (Mills-Bria et al., 2006)

Reclamation facility tested 6-inch diameter cores (i.e. with a maximum aggregate size of approx. 6 inches) drilled perpendicular to the top surface dams and sealed in plastic to preserve the in situ moisture content during shipping. Table lists the average, maximum and minimum values for eleven dams. The ratio of the maximum average to

TABLE 6.1

Summary of static compression tests performed on 11 dams (Mills-Bria et al., 2006)

Project	Average			Maximum			Minimum		
	Strength MPa	E GPa	Strain $\times 10^{-6}$	Strength MPA	E GPa	Strain $\times 10^{-6}$	Strength MPa	E GPa	Strain $\times 10^{-6}$
Deadwood	32.4	24.1	1,785	41.5	29.7	2,240	22.1	20.7	1,103
Elephant Butte	17.4	19.1	1,450	31.1	30.3	2,055	8.8	10.3	
Englebright	45.0	32.4		45.0	32.4				
Folsom	29.3	14.5		29.3	14.5				
Hoover	47.5	38.6		63.6	51.7		29.3	14.5	
Monticello	30.5	35.6	1,183	40.1	49.6	1,400	41.9	26.9	
Pine Flat	26.8	26.9		26.8	26.9				
Roosevelt	37.3	37.9	1,175	48.8	55,2	1,625	28.2	18.6	
Seminoe	24.1	11.7	951	36.4	22.4	2,880	15.6	6.9	
Stewart Mountain	34.8	26.9		46.0	40.0	24.8	14.5		
Warm Springs	20.2	23.4		46.0	46.2		10.2	5.5	
All projects	31.4	26.2		63.6	55.1		8.8	5.5	

the minimum average is observed to be approximately 7:1 for compressive strength, 10:1 for the modulus of elasticity and 4:1 for compressive strain at failure. If examined project-by-project, the maximum range would be 3.5 times for compressive strength, 8 times for modulus of elasticity and 2 times for compressive strain at failure.

It thus appears obvious that the properties for a given structure should be based on multiple tests on cores retrieved from the altered structure.

Moreover, due to uncertainly regarding project-specific materials, the properties selected for input should include a bandwidth of $\pm\sigma\%$ of expected values, where σ depends on the number of tests performed. If no more than one test data point were available, then σ should equal 15-20%.

6.1.2 UNITS & CONVERSION FACTORS

Given the multiple units included among the various physical parameters in an integrated AAR study, Table 6.2 provides convenient conversion factors for parameters required in the numerical simulation of AAR.

TABLE 6.2

Conversion Factors

Mechanics		
Length, m (meters)	1 inch = 0.0254 m	1 m = 39.37 inch
Force, N (Newtons)	1 lb = 4.4482 N	1 N = 0.22481 lb
Mass, kg (kilograms)	1 lbm = 0.45359 Kg	1 Kg=2.2046 lb
Density, kg/m^3	1 lbm/ft^3 = 16.018 Kg/m^3	1 Kg/m^3=0.062428 lbm/ft^3
Temperature, T	T $^\circ$F=[(9/5)T$^\circ$C+32]	
Acceleration, m/s^2	1 in/s^2 = 0.0254 m/s^2	
Stiffness, N/m	1 lb/in = 175.1 N/m	
Stress, Pa = N/m^2	1 psi = 6,894.8 Pa	1 MPa = 145.04 psi
Work, energy, N-m=Joule	1 ft-lbf= 1.3558 J	1 J = 0.73756 ft- lbf
Heat Transfer		
Convection coefficient, h	1 Btu.h.ft^2.$^\circ$F = 5.6783 W/m^2.$^\circ$C	
Heat, J	1 Btu=1055.06 J	1 Btu = 778.17 ft-lb
Heat source/sink, Q	W/m^3 =	
Heat flux (q)	1 Btu/h.ft^2 = 3.1546 W/m^2	
Specific heat, c	1 Btu/$^\circ$F = 1,899.108 J/$^\circ$C	
Thermal conductivity, k	1 Btu/h.ft.$^\circ$F = 1.7307 W/m.$^\circ$C	
Fracture Mechanics		
Stress intensity factor, K	1 MPa$\sqrt{m}$=1.099 ksi$\sqrt{in}$	
Fracture energy G_F	1 lb/in =.0057 N/m	

6.2 ELASTIC PROPERTIES

Elastic properties are the most basic mechanical properties needed to conduct a finite element analysis; this section will individually address each of the most essential elastic properties. We will omit compressive strength since it is the unanimously tested property and most often serves to derive the other quantities (similar to the physical quantities expressed in terms of L, T and M for length, time and mass, respectively).

This section will not address the nonlinear concrete properties discussed in Sect. 3.2

6.2.1 ELASTIC MODULUS

Based on a large number of tests conducted, ACI 318 (2011) has established the following empirical relationship between compressive strength f_c' MPa (lb/in^2) and the unit weight of the concrete kg/m^3 (lb/ft^3):

$$E_c = \begin{cases} 0.043 w_c^{1.5} \sqrt{f_c'} & \text{MPa} \\ 33 w_c^{1.5} \sqrt{f_c'} & \text{lb/in}^2 \end{cases} \tag{6.1}$$

Assuming a density of normal weight concrete equal to 2,320 kg/m^3 (145 lb/ft^3), this expression can be reduced to:

$$E_c = \begin{cases} 4,730 \sqrt{f_c'} & \text{MPa} \\ 57,000 \sqrt{f_c'} & \text{lb/in}^2 \end{cases} \tag{6.2}$$

It is commonly assumed that the modulus of elasticity remains the same whether in tension or compression.

6.2.2 TENSILE STRENGTH

The direct tensile strength of concrete is seldom determined since the corresponding determination step requires a non-codified testing apparatus, such as the set-up shown in Figure 6.2. Should laboratory tests be unavailable, empirical equations can be used to estimate concrete tensile strength values.

Empirical relations for tensile strength are given by:

$$f_t' = \begin{cases} 3 - 5\sqrt{f_c'} & \text{psi} \quad \text{ACI-318} \\ 1.4 \left(\frac{f_{ck}}{f_{ck0}}\right)^{2/3} & \text{MPa} \quad \text{CEB-FIP} \end{cases} \tag{6.3}$$

where: f_{ck} is the cylinder compressive strength and f_{ck0} equals 10 MPa.

For mass concrete, Raphael (1984) studied concrete tensile strength from nearly 12,000 concrete specimens and showed the limited basis for assuming a linear relationship between tensile and compressive strength. The author compared: 1) the relationship between direct tensile strength, splitting tensile strength and modulus of rupture; 2) the relationship between linear and nonlinear assumptions applicable to concrete strength; and 3) the relationship between the static and dynamic strength of concrete. It was found that the direct tensile strength is typically about half the splitting tensile strength. It was also demonstrated that the drying of cylinders affects their strength.

In the absence of test results, Raphael suggested a method for estimating splitting strength from the compressive strength of concrete, namely:

$$f_{st} = \frac{2P}{\pi LD} = \alpha f_c^{2/3} \tag{6.4}$$

where:

α 1.7 for static loading
       2.3 for static loading, though this accounts for concrete nonlinearity
       2.6 for seismic loading
f_{st} Splitting tensile strength
P Compressive force applied to the specimen
L Specimen length
D Specimen diameter
f_c Uniaxial static compressive strength of the concrete

Raphael also recommended using $\alpha = 2.3$ for static finite element analysis and:

$$f_{st} = \frac{2P}{\pi LD} = 3.4 f_c^{3/2} \tag{6.5}$$

for linear finite element analysis under seismic loading.
     The modulus of rupture can then be estimated from:

$$f_{mr} = \frac{PL}{bd^2} = 1.33 f_{st} = 2.3 f_c^{2/3} \tag{6.6}$$

where:

f_{mr} Tensile strength obtained from a modulus of rupture test
P Force applied to the specimen
L Specimen length

(a) Experimental set-up   (b) Ruptured specimen

FIGURE 6.2   Experimental set-up for a direct tension test (Delft University)

b Specimen width
d Specimen depth
f_c Uniaxial static compressive strength of the concrete

6.2.3 POISSON'S RATIO

The Poisson's ratio for concrete typically lies between 0.2 and 0.25. Results from the structural analysis are relatively insensitive to the assumed value. As a consequence, Poisson's ratio is typically set to 0.2 (unless data are available).

6.2.4 FRACTURE PROPERTIES

Fracture properties are discussed separately in Sect. 3.2.2.1

6.3 AAR PROPERTIES

Once again and in accordance with the model presented by the author, the two most important parameters are the activation energies for both characteristic and latency times. These values can be determined experimentally through Eq. 2.6 (and the accompanying Fig. 2.1(b)); however, it has been indicated that these are indeed universal constants[1]. Both Larive (1998a) and Ben Haha (2006) derived sufficiently close values, such that we can assume Eq. 2.7 is applicable with:

$$
\begin{aligned}
U_l &= \frac{E_l}{R} = 9,400 \pm 500K \\
U_c &= \frac{E_c}{R} = 5,400 \pm 500K
\end{aligned}
\tag{6.7}
$$

It should be noted that activation energies (E_l and E_c) are expressed in units of J/M, while the universal gas constant R is in J/M.K (8.31).

The latency time, characteristic time and maximum AAR strain (τ_l, τ_c and ε^∞) can only be determined from (accelerated) core expansion tests. As previously shown in Table 2.1, substantial variability can also be present for the same concrete specimen due to the fact that not all aggregates are equally reactive in a concrete mix.

6.4 THERMAL PROPERTIES

The importance of adequate temporal and spatial distribution in an AAR analysis has been emphasized often. Section 1.4 presented a brief review of thermal analysis; this section is intended to simply provide assistance in assigning (air) temperature and physical property values as part of a thermal analysis.

---

[1]Let's note that Shon (2008) has determined that activation energy is a function of aggregate reactivity, i.e. activation energy increases with reactivity. These results however do not appear to be well substantiated.

6.4.1 TEMPERATURES

6.4.1.1 Air temperature

The daily variation in air temperature obviously depends on the geographic location. The National Oceanographic and Atmospheric Agency (NOAA) has set up a National Climatic Data Center http://www.ncdc.noaa.gov/oa/climate/climatedata.html# surfacehttp://www.ncdc.noaa.gov/oa/climate/climatedata.html#surface containing thousands of locations worldwide that can be accessed for thermal analyses.

On the other hand, should data be limited, the transient variation of temperature can be estimated based on T_{mean}, T_{min} and T_{max}, which denote the mean, minimum and maximum temperatures respectively:

$$T_a(t) = A \sin\left(\frac{2\pi(t - \xi)}{365}\right) + T_{mean} \tag{6.8}$$

where $A = 0.5\left(|T_{max} - T_{mean}| + |T_{min} - T_{mean}|\right)$, t is the time in days, and ξ the time in days when $T_a = T_{mean}$. To avoid having to use the highly nonlinear equation for temperature exchange through radiation (Eq. 1.40), a common approach is to increase ambient temperature by 0.5 to 1.0 deg C (i.e. 1 to 2 deg F). USACE (1994) and ACI-207 (2005) provided charts to allow approximating the estimates of solar radiation effects.

6.4.1.2 Pool temperature

An empirical equation for the reservoir temperature in terms of depth has been given by Anon. (1985):

$$T_w(y) = c + (T_{surf} - c)\, e^{-0.04y} \tag{6.9}$$

where:

$$c = \frac{T_{bot} - e^{-0/.04H} T_{surf}}{1 - e^{-0.04H}} \tag{6.10}$$

and T_{bot} T_{surf}, H and y are respectively the annual mean water temperatures at the bottom and surface, the reservoir depth (in meters), and the water depth in meters.

6.4.2 CONCRETE THERMAL PROPERTIES

The material properties required to conduct a thermal analysis are summarized in Table 6.3 Thermal diffusivity h is a measure of the rate at which temperature change can occur in a material; its value is derived by dividing the thermal conductivity by the product of specific heat and unit weight. For mass concrete, typical thermal diffusivity values range from 0.003 to 0.006 m^2/hr.

For mass concrete, the thermal conductivity k equals approx. 2.7 J/sec.m.K., and the specific heat varies between 0.75 and 1.17 kJ/kg-K.

The film coefficients for heat transfer by convection are on the order of $h_{air} = 34 W/m^2 \ ^oC$ and $h_{water} = 100 W/m^2 \ ^oC$.

TABLE 6.3

Material parameters required for a thermal analysis

	Steady-state	Transient
Material properties		
Mass density		ρ
Specific heat		c
Conductivity	k	k
Boundary conditions		
Temperature	T	T
Film	h	h
Flux	q	q

It should be noted that in a stress analysis, results are very sensitive to the selected coefficient of thermal expansion α, which depends on the type of aggregate and can be assumed equal to 1×10^{-5} m/m/°C.

6.5 RECLAMATION STUDY

6.5.1 ELASTIC PROPERTIES

As pointed out earlier (Section 1.2.2), the reaction is accompanied by a degradation of both the elastic modulus and tensile strength. In the absence of test results from recovered cores, it may be worthwhile to consult the extensive database assembled by the Bureau of Reclamation on properties of old dams measured over many years. This source could serve as an excellent starting point for preliminary analysis. This section will provide an adaption of Dolen (2005, 2011) and the text/figures displayed herein have been reproduced with the permission of the Bureau of Reclamation.

A database model of aging concrete was developed by the Bureau of Reclamation in order to identify the changes in material properties over time. Data on mass concrete material properties were input into the *Aging Concrete Information System* (ACIS), where they were analyzed for trends in the deterioration of concretes subject to aging, including the alkali aggregate reaction (AAR) and general aging of early 20th-century concrete with high water-to-cement ratios.

These aging concrete samples were then compared to dams from a similar period, yet without any adverse effects from aging processes. The aging concretes were also compared to known high quality concretes mixed after circa 1948 specifically to resist deterioration from AAR, freezing and thawing (FT), and sulfate attack. Trends were established so as to compare the compressive strength, splitting and direct tensile strength, and elastic properties of both aging and non-aging dams.

The strength and elastic properties of aging mass concrete differed significantly from those of comparable non-aging specimens. Both the spatial variations within a structure and long-term changes in strength and elastic properties could be identified.

Three Bureau of Reclamation concrete dams have sustained significant deterioration attributable to AAR, and in particular ASR. Parker Dam, built in 1937-38, was the first Reclamation dam to be identified with ASR. American Falls Dam, which dates back to 1927, was actually the first Reclamation facility to experience deterioration caused by ASR and was ultimately replaced in 1977. Seminoe Dam was constructed in 1938 and has gradually deteriorated over time. Both Parker and Seminoe Dams feature comparable "reference" dams, i.e. built using similar materials and mixtures at about the same time with little deterioration.

6.5.1.1 Effect of confinement

Confinement plays an important role in reducing the elastic properties of concrete. Samples taken above 20 feet differ markedly from those extracted below the 20-foot level due to the (beneficial) effect of confinement. Table 6.4 clearly illustrates this effect. These data are to be compared with the non-ASR-affected concrete modulus

TABLE 6.4

Average compressive strength and elastic properties of concrete dams subject to aging (Dolen, 2005)

	Test Age years	Compressive strength lb/in^2	Modulus of Elasticity 10^6 lb/in^2	Poisson's ratio
AAR-affected dams*	53.1	3,695	2.28	0.20
AAR cores from above 20 ft	48.0	3,180	2.09	0.20
AAR cores from below 20 ft	49.0	4,090	2.35	0.10
ACIS aging dams (1902 to 1920)*	79.7	2,490	2.59	0.23

*The average is weighted by the number of tests for a given sample set.

value of approx. 5.410^6 lb/in^2.

6.5.2 COMPRESSIVE STRENGTH

Figure 6.3 shows the relationship between compressive strength and modulus of elasticity for all concrete cores both with and without ASR. Although the correlation coefficient for these equations is poor, the trend lines have been added to indicate the demarcation between the two classes of concrete. Individual correlations between compressive strength and modulus of elasticity are typically much better for individual dams using the same types of aggregate. These trends reveal that the strength-to-modulus of elasticity relationship provides a good indicator of ASR and may be used to develop failure criteria for predictive models.

6.5.3 TENSILE STRENGTH

Tensile strength is becoming more critical in the structural analysis of concrete dams, particularly in the case of dynamic analyses involving earthquakes. Tensile strength tests were not normally performed until the 1970's, and tensile strength development in dams built before this era is unknown. The results of direct and splitting tensile strength tests on high quality concrete and aging/ASR-affected concrete specimens are shown in Figure 6.4. Tensile strength data have typically been entered into the database as average values from just a few tests for each mixture. The aging data also include some tests on old dams not subject to ASR. It is clear however that the tensile strength of aging/ASR-affected dams is averaged at about 50 percent of the direct tensile strength and 30 percent less in splitting tensile strength, when compared to dams without ASR degradation or aging. Also, aging concrete data are often based solely on âŁœtestableâŁž concrete and do not represent the condition of the deteriorated concrete that could not be tested. Lift line ratios may not be directly comparable since aging dams often contain more disbonded lift lines. This input parameter has been added to more recent testing programs and constitutes a factor for some older and newer dams. Shear bond properties are not included in this dataset and moreover have not yet been analyzed due to insufficient records.

6.5.4 CASE STUDIES

The evolution in strength at both Parker and Seminoe Dams has been extensively studied for ASR effects. Cracking at Parker Dam was identified as ASR after examinations confirmed the process first identified by Stanton (1940). Widespread concrete

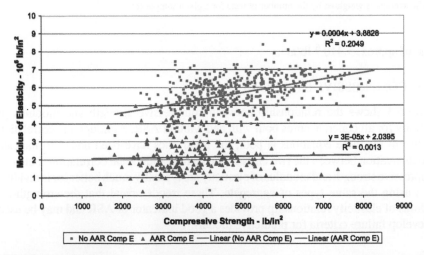

FIGURE 6.3 Comparison of strength with modulus of elasticity in compression for mass concrete dams both with and without ASR (Dolen, 2005)

coring and testing have been performed since 1940, and the resulting strength trends are well documented. Seminoe Dam was subjected to early freezing and thawing near the dam crest, but ASR was not identified as a significant contributing factor to degradation until more than 60 years after construction. The deterioration at Seminoe Dam seems more alarming because the mass concrete appears to have nearly reached its projected ultimate strength potential before the onset of ASR. The slow reaction rate may be due to both the nature of the aggregates and the cold temperatures at the site. Another dam built in a northern climate dam with the potential for similar behavior is Owyhee Dam in Idaho. Tests conducted near the crest of Owyhee Dam reveal a behavior close to that found at Seminoe Dam, and potentially reactive aggregates are indeed prevalent in the vicinity of Owyhee Dam.

The compressive strength trend lines in Figure 6.5 from the two ASR-affected dams show a relatively constant state for Parker Dam and a decreasing strength at Seminoe Dam. Some of the data scatter is caused by the fact that overall sampling has not been sorted by elevation and moreover includes test results on concrete not significantly affected by ASR due either to the cement alkali content of individual block placements or to location in the dam. When sorted by elevation, the rate of change can also be observed for both dams, as depicted in Figures 6.6 and 6.7. These compressive strength trends do not indicate an overall change with time for Parker Dam, even though some spatial trends may be present. For Seminoe Dam, it is readily apparent that the overall compressive strength is decreasing with time and the compressive strength exhibits significant spatial deterioration near the top of the dam, as shown in Figure 6.8. This deterioration extends more deeply into the dam over time. The modulus of elasticity also displays the same trends (Fig. 6.9).

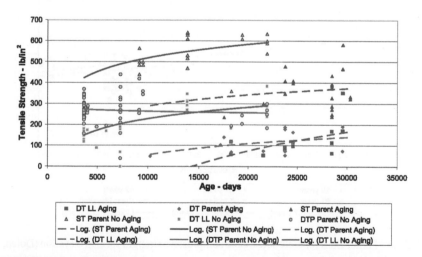

FIGURE 6.4 Comparison of the effects of aging and ASR on the tensile strength of mass concrete dams (Dolen, 2005)

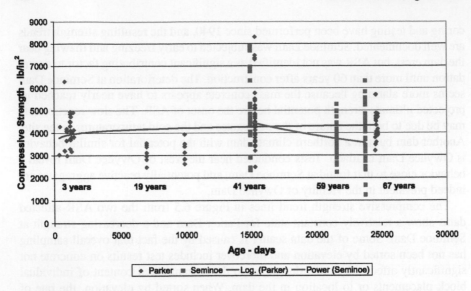

FIGURE 6.5   Compressive strength development in ASR-affected concrete cores for Parker and Seminoe Dams (Dolen, 2005)

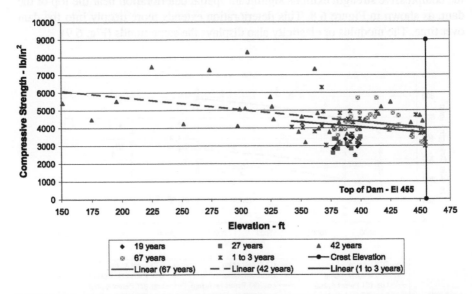

FIGURE 6.6   Parker Dam concrete cores; Core compressive strength vs. dam elevation (Dolen, 2005)

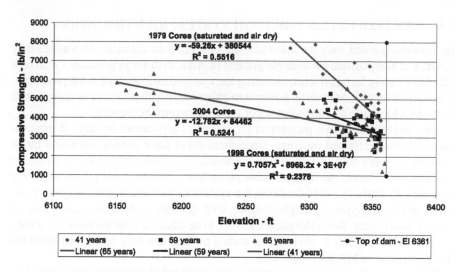

FIGURE 6.7   Effect of both ASR reaction and freezing & thawing; Seminoe Dam; Compressive strength vs. elevation (Dolen, 2005)

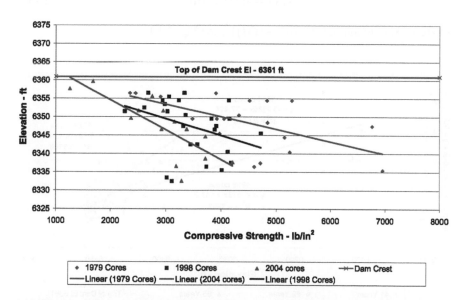

FIGURE 6.8   Effect of ASR reaction and freezing & thawing; Top 30 feet of Seminoe Dam; Compressive strength vs. elevation (Dolen, 2005)

6.6   AAR PROPERTIES THROUGH SYSTEM IDENTIFICATION

As previously noted, only the selected zones of a structure are most often affected by AAR; it is also possible for multiple pockets of AAR to differ in intensity. Great care should therefore be taken when assessing the spatial distribution of AAR properties. It is seldom feasible or realistic however to either identify these zones or characterize them, in which case an equivalent set of parameters must be assumed.

The three critical AAR parameters (τ_l, τ_c and $\varepsilon(\infty)$) cannot be derived from expansion tests performed on cores, since these cores have already expanded by an unknown amount once they have been retrieved. This challenging issue is discussed in Chapter 9. Hence, the only effective procedure to derive AAR parameters for a time-history analysis would be through a system identification that compares measured quantities (typically displacements) with quantities obtained numerically. In case of discrepancy, the model parameters are to be adjusted until the error lies within an acceptable range. It is obvious that such a method can only be applied provided the availability of historical field data.

Mathematically speaking, the problem can be simply formulated as follows. The field-recorded displacements (e.g. crest displacement on a dam) are denoted by $\mathbf{u}(t)$, the target parameters by $\mathbf{x}$ (in our case $x(1) = \tau_c$, $x(2) = \tau_l$ and ($x(3) = \varepsilon(\infty)$), the finite element "operator" by $f(.)$, and computed results by $\mathbf{u}'(t)$. We thus have:

$$f(\mathbf{x}) = \mathbf{u}'(t) \neq \mathbf{u}(t) \tag{6.11}$$

and are seeking to minimize $(\mathbf{u}(t) - \mathbf{u}'(t))^2$, see Fig. 6.10.

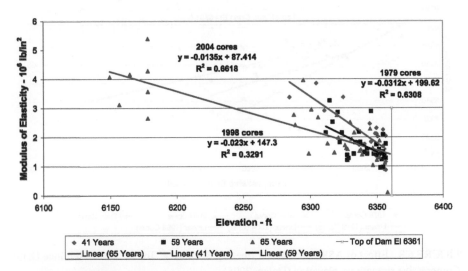

FIGURE 6.9   Effect of ASR reaction and freezing & thawing; Seminoe Dam; Modulus of elasticity vs. elevation (Dolen, 2005)

The corresponding objective function, Jacobian, gradient and Hessian matrices Dennis and Schnabel (1983) are then given by:

$$\omega(\mathbf{x}) \;=\; (\mathbf{u}-\mathbf{u}')^T(\mathbf{u}-\mathbf{u}') = \mathbf{r}^T \cdot \mathbf{r} = \sum_{i=1}^{m}(u_i - u_i')^2 \qquad (6.12)$$

$$\mathbf{J}(\mathbf{x}) \;=\; \frac{\partial \mathbf{r}(\mathbf{x})}{\partial \mathbf{x}^T} = \frac{\partial}{\partial \mathbf{x}^T}\left[(\mathbf{u}-\mathbf{u}')\right] = \frac{\partial \mathbf{u}}{\partial \mathbf{x}^T} - \frac{\partial \mathbf{u}'}{\partial \mathbf{x}^T} = -\frac{\partial \mathbf{u}'}{\partial \mathbf{x}^T} = -\mathbf{L} \quad (6.13)$$

$$\nabla\omega(\mathbf{x}) \;=\; \frac{\partial \omega(\mathbf{x})}{\partial \mathbf{x}^T} = \frac{\partial}{\partial \mathbf{x}^T}\left[(\mathbf{u}-\mathbf{u}')^T(\mathbf{u}-\mathbf{u}')\right] = 2\mathbf{J}^T\mathbf{r} \qquad (6.14)$$

$$\mathbf{H}(\mathbf{x}) \;=\; \frac{\partial^2 \omega}{\partial \mathbf{x}^T \partial \mathbf{x}} = \frac{\partial}{\partial \mathbf{x}}\left(\nabla\omega(\mathbf{x})\right) = \frac{\partial}{\partial \mathbf{x}}\left[2\mathbf{J}^T\mathbf{r}\right] = 2\mathbf{J}^T\mathbf{J} + 2\frac{\partial^2 \mathbf{r}}{\partial \mathbf{x}^T \partial \mathbf{x}}\mathbf{r} \;(6.15)$$

In nearing the minimum, $\frac{\partial^2 \mathbf{r}}{\partial \mathbf{x}^T \partial \mathbf{x}}\mathbf{r} \to 0$, we have $\mathbf{H}(\mathbf{x}) = 2\mathbf{J}^T\mathbf{J}$. Use of the Levenberg-Marquardt method Dennis and Schnabel (1983) allows solving for $\mathbf{x}_{k+1}$ from:

$$\mathbf{x}_{k+1} = \mathbf{x}_k - \left[\epsilon_k \mathbf{I} + \mathbf{H}(\mathbf{x}_k)\right]^{-1}\nabla\omega(\mathbf{x}_k) \qquad (6.16)$$

where ϵ_k is such that all eigenvalues of $[\epsilon_k\mathbf{I} + \mathbf{H}(\mathbf{x}_k)]$ are positive and greater than $\delta > 0$. A development of the Levenberg-Marquardt method, known as the trust region method, can ultimately be introduced. We have indeed sought to minimize the objective function $\omega(\mathbf{x})$ inside a trust region, where the quadratic approximation (as obtained by a Taylor series) is considered reliable:

$$\omega_{II}(\mathbf{x}, \mathbf{x}_k) = \omega(\mathbf{x}_k) + \nabla\omega(\mathbf{x}_k)^T(\mathbf{x}-\mathbf{x}_k) + \frac{1}{2}(\mathbf{x}-\mathbf{x}_k)^T\mathbf{H}(\mathbf{x}_k)(\mathbf{x}-\mathbf{x}_k)$$
$$(6.17)$$

The trust region is defined as follows:

$$\Omega_k = \{\mathbf{x} : \|\mathbf{x}-\mathbf{x}_k\| \leqslant \Delta_k\} ; \Delta_k > 0 \qquad (6.18)$$

and the optimization problem becomes:

$$\min_{\mathbf{x}}\{\omega_{II}(\mathbf{x}, \mathbf{x}_k) : \mathbf{x}_k \in \Omega_k\} \qquad (6.19)$$

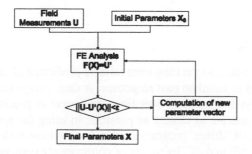

FIGURE 6.10   Principle of the system identification approach

The condition needed to update the region is given by:

$$R_k = \frac{\omega\left(\mathbf{x}_k\right) - \omega\left(\mathbf{x}_{k+1}\right)}{\omega\left(\mathbf{x}_k\right) - \omega_{II}\left(\mathbf{x}_{k+1}\right)} \simeq 1 \tag{6.20}$$

The update of the starting point is the same as that illustrated for the Levenberg-Marquardt method. With this method, we are also able to define the upper and lower bounds for these parameters as:

$$\min_{\mathbf{x}} \left\{\omega_{II}\left(\mathbf{x}, \mathbf{x}_k\right) : \mathbf{x}_k \in \Omega_k\right\} ; \text{lb} \leqslant \mathbf{x} \leqslant \text{ub} \tag{6.21}$$

From a practical standpoint, the three following considerations must be taken into account during the optimization process:

The starting point of the parameter identification process may be determined from laboratory tests (yielding an initial estimate of $\varepsilon^\infty|_{T=T_0}$, $\tau_L(T_0)$ and $\tau_C(T_0)$). Alternatively, curve fitting of the dam crest displacement:

$$u_{irr}^{AAR}(t, \theta) = \frac{1 - e^{-\frac{t}{\tau_c(\theta)}}}{1 + e^{-\frac{(t-\tau_l(\theta))}{\tau_c(\theta)}}} u_{irr}^{AAR,\infty} \tag{6.22}$$

(where u_{irr}^{AAR} and $u_{irr}^{AAR,\infty}$ are the irreversible, time and temperature-dependent displacement and the final displacement due to AAR respectively) provides a conservative estimate for the times, but not for $\varepsilon^\infty|_{T=T_0}$. The target parameters must be normalized such that all initial values have the same order of magnitude, i.e.: a) the initial variation must be large enough to produce a variation in the computed results (large normalized values can lead to an immediate stop of the identification process); and b) the final variation of parameters must be small enough to allow for a small final adjustment of the identified parameters without large oscillations around the final solution (an overly small normalized parameter may result in a large final oscillation).

A weight function can be introduced to assign importance to the last data field, which typically contains a major absolute value and is thus more representative of the irreversible effect of AAR expansion with respect to the effect of normal loads. Let's note that this system identification does not require a thermal analysis and, from a practical standpoint, entails a simple modification of three variables included in the stress analysis input file.

6.6.1 ALGORITHM

Of paramount importance to the long-term reliable prediction of dam response is the ability of the model to simulate past responses. A dam always contains a record of the irreversible crest displacement over time; this curve is precisely what a model should be able to replicate as closely as possible. In using the approach previously described, a MATLAB "driver" program has thus been written to identify the three key parameters $\tau_l(\theta)$, $\tau_c(\theta)$ and $\varepsilon_\theta^\infty$, by means of continuously comparing the histogram of the computed crest displacement with the histogram of recordings. The driver will:

1. Launch an initial finite element analysis.
2. Open the output file, extract the crest displacement histogram.
3. Compare the numerical histogram with field recordings. If the results are nearly similar, stop execution.
4. Appropriately modify the input key parameters, and launch the finite element analysis.

This procedure is best illustrated by Figure 6.11 where: Left: Matlab code; upper

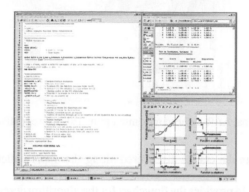

FIGURE 6.11   Graphical user interface for AAR parameter identification

right: Finite Element analysis program execution; lower right: Comparison of computed and recorded crest displacements (upper left), and Variation of τ_l, τ_c and ε^∞ vs. number of iterations.

Figure 6.12 shows that the computed crest displacement lies well within the sea-

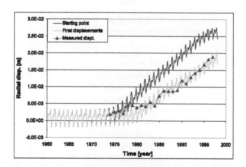

FIGURE 6.12   Comparison between the first analysis estimate and the final analysis using optimized parameters

sonal variations of the final numerical predictions.

For a 3D mesh (7,552 nodes and 5,196 elements), the total identification procedure lasted about 190 hours on a Pentium IV computer. Figure 6.13 presents the internal AAR-induced maximum principal stresses. The maximum principal stress field inside

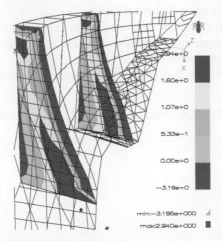

FIGURE 6.13   Internal AAR-induced maximum principal stresses (MPa)

the dam can explain the cracks discovered along the upper gallery of the analyzed dam.

Lastly, Figure 6.14 provides the flowchart for the algorithm described above. Below is the Matlab listing for the proposed algorithm.

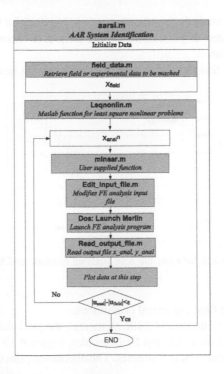

FIGURE 6.14   AARSI flowchart

Listing 6.1   AARSI.m

```
 1  %==================================================
 2  %==================================================
 3  % AARSI.m
 4  % Alkali Aggregate Reaction System Identification
 5  %==================================================
 6  %==================================================
 7  %==================================================
 8  % Global declaration
 9  %==================================================
10  clear global
11  clear              % Clear all arrays
12  clf                % Clear figure
13  global kplot x_exp y_exp n_x_anal y_anal n_increments n_iterations n_parameters ...
14      factor history flag_program var
15  %-------------------START------------------------'
16  %==================================================
17  % Initialization
18  %==================================================
19  maxfuneval=10000;     % Maximum function evaluation
20  maxiter=1000;         % Maximum iterations
21  tolfun=1e-8;          % Tolerance for the function
22  tolx=1e-8;            % Tolerance for the unknowns
23  maxPCGiter=100;       % Maximum number of PCG iterations
24  tolPCG=0.00001;       % Termination tolerance on the PCG iteration
25  t_start=cputime;      % Set the chronometer to
26  % x(1)                % epsi_infty
27  % x(2)                % Characteristic time
28  % x(3)                % Latency time
29  % x(4)                % Activation energy for characteristic time
30  % x(5)                % Activation energy for latency time
31  % x(6)                % Fraction of compressive strain (AAR expansion diminution)
32  % x(7)                % Fraction of tensile strain (AAR expansion diminution)
33  % x(8)                % Critical compressive strain (Not failure)
34  % x(9)                % Critical tensile strain (Not failure)
35  % x(10)               % Curve weight in compression
36  % x(11)               % Reference temperature
37  % x(12)               % Stress for zero AAR strain
38  % ub_default          % Default upper bound vector
39  % lb_default          % Default lower bound vector
40  % merlin.inp          % Name of Merlin input file
41  % merlin.out          % Name of Merlin output file
42  % x_anal, y_anal      % Numerically computed x and y to be matched with:
43  % x_exp, y_exp        % x and of field/lab data to be matched
44
45  %==================================================
46  % Retrieve experimental data
47  %==================================================
48  '--------- RETRIEVING EXPERIMENTAL DATA----------';
49  field_data;
50  %==================================================
51  % plot initial experimental data
52  %==================================================
53  %subplot(2,2,1); kplot=plot(x_exp,y_exp,'-r','linewidth',2); % Label the plot to later modify it
54  %xlabel('Time'); ylabel('Longitudinal strain');
55  %grid
56  %hold on
57  %==================================================
58  % Initialize variables, lower and upper bounds
59  %==================================================
60  % Note that variables are normalized such that
61  flag_program=[1;1;1;0;0;0;0;0;0;0;0;0];
62  factor_default=[1e4;1e9;1e9;0;0;5e5;0;-1e3;0;1e8;0;0];
63  ub_default=[1e-5;1e-5;1e-5;1e-5;1e-5;1e-6;1e-5;1e-5;1e-6;1e-5;1e-5;1e-5];
64  lb_default=[0;0;0;0;0;0;0;0;0;0;0;0];
65  vet_x_default=[1e-7;1e-7;1e-7;1e-7;1e-7;1e-7;1e-7;1e-7;1e-7;1.5e-7;1e-7;1e-7];
66
67  k=0;
68  for i=1:12
69      if flag_program(i)==1;
70          k=k+1;
71          vet_x(k,1)=vet_x_default(i);
72          factor(1,k)=factor_default(i);
73          vet_ub(k,1)=ub_default(i);
74          vet_lb(k,1)=lb_default(i);
75      end
76  end
77  n_parameters=k;
78  xo=vet_x;
79  ub=vet_ub;
80  lb=vet_lb;
81  t_start=cputime;
82  kount=0;
83  %==================================================
84  % Start optimization
85  %==================================================
86  %==================================================
```

```
 87   n_iterations=0;
 88   %========================================
 89   % set options for the optimization function
 90   options=optimset('Display','iter','Jacobian','off','MaxFunEvals',...
 91       maxfuneval,'Maxiter',maxiter,'TolFun',tolfun,'TolX',tolx,...
 92       'LargeScale','on','LevenbergMarquardt','off','MAxPCGIter',...
 93       maxPCGiter,'TolPCG',tolPCG);
 94   % input the nonlinear minimization function
 95   [x,resnorm,residual,exitflag,output,jacobian]=lsqnonlin(@minaar,xo,lb,ub,options);
 96   t_end=cputime;
 97   t_total=t_end-t_start;
 98   %save converged data on file
 99   fn='Hmat.out';
100   fstor=fopen(fn,'r+');
101   frewind(fstor);
102   fprintf(fstor,'%16.12e \n',history);
103   fclose(fstor);
104   %----------------------END---------------------';
```

Listing 6.2   minaar.m

```
 1   %========================================
 2   %========================================
 3   % function minaar
 4   % Computes the objective function (residual) to be minimized
 5   %========================================
 6   %========================================
 7   function [e]=minaar(x)
 8   %========================================
 9   %% Global declaration
10   %========================================
11   global kplot x_exp y_exp x_anal y_anal n_increments history n_iterations ...
12       flag_program n_parameters
13   %========================================
14   %% Update Merlin's input with current parameters
15   %========================================
16   %-----------EDIT INPUT DATA FILE-----------';
17   edit_input_file;
18   %========================================
19   %% Start up analysis
20   %========================================
21   %-----------LAUNCH MERLIN ANALYSIS-----------';
22   status=dos ('c:\Dam_Software\Release_bin\merlin merlin.inp merlin.out merlin.pst');
23   %========================================
24   %% Extract results
25   %========================================
26   %-----EXTRACT RESULTS FROM MERLIN ANALYSIS-----';
27   Read_output_file;
28   %========================================
29   %% determine the residual of displacements
30   %========================================
31   residual=((y_anal(2:n_increments)-y_exp(2:n_increments))./...
32       y_exp(2:n_increments));
33   e=residual;
34   norm_e=norm(e);
35   str1='Norm=';
36   str2=num2str(norm_e,'%9.2e');
37   string=strcat(str1,str2);
38   %========================================
39   %% Get min and max of y
40   %========================================
41   y_min=min(min(y_anal,y_exp));y_max=max(max(y_anal,y_exp));
42   %========================================
43   %% plot results
44   %========================================
45   %Iteration number
46   for k=1:n_iterations
47       vet_iterations(1,k)=k;
48   end
49   %Subplot
50   column=2;
51   i=n_parameters;
52   %====
53   row=0;
54   while i>=0
55       i=i-2;
56       row=row+1;
57   end
58   %====
59   ff=1;
60   text=['AAR strain        ';'Charac time       ';'Latency time      ';...
61       'Act energy tc     ';'Act energy tl     ';'Gamma c           ';...
62       'Gamma t           ';'Crit comp strain  ';'Crit tens strain  ';...
63       'Curve weight      ';'Ref temperature   ';'Sigma 2           '];
```

```
64   for f=1:12;
65       if flag_program(f,1)==1
66           text1(ff,:)=text(f,:);
67           ff=ff+1;
68       end
69   end
70   %_____
71   % Plot initial experimental data
72   %_____
73   subplot(row,column,1); plot(x_anal,y_anal,'-k',x_exp,y_exp,'-r','linewidth',2);
74   xlabel('Time'); ylabel('Longitudinal strain');
75   grid
76   %hold on
77   %_____
78   % Plot convergence data
79   %_____
80   for j=1:n_parameters
81       subplot(row,column,j+1); plot(vet_iterations,history(:,j),'-ok');
82       xlabel ('Function evaluations');
83       ylabel (text1(j,:));
84       grid
85   end
```

6.7   ON THE IMPORTANCE OF PROPER CALIBRATION

All too often, models are solely calibrated with displacements and seldom with stresses. Though displacements may occasionally be significant in a structure, strains and stresses are more critical (and themselves related to the derivative of displacement). In the following example, we will consider a reference (as an arbitrarily-defined analytical) curve u_{ref} and a model response u_{model}. Figure 6.15(a) shows a nearly perfect match between the two, suggesting a potentially satisfactory model prediction. However, if the derivatives of these two curves were to be plotted, see Fig. 6.15(b), it becomes obvious that the error is much more substantial, as highlighted in Figure 6.15(c), which displays the percentage error in each.

Measuring in situ stresses is not a straightforward step, as either flat-discs (limited to surface measurements), over-coring or stress-meters may be used, Fig. 6.16. Most current in situ stress measurements in dams are performed using the over-coring technique. Yet this methodology can prove to be expensive and moreover does not allow for repeated readings at the same location/borehole.

In contrast, the stress-meter can output repeated readings of the in situ elastic modulus and stresses through a probe yet is only seldom used. The author has worked with a stress-meter during a past project on a dam (Saouma et al., 1991a) and strongly recommends its adoption in monitoring AAR progress over time through repeated readings and measurement of (degrading) elastic properties, along with variations in the magnitude/direction of in situ stresses.

6.8   SUMMARY

One can never overemphasize enough the importance of credible material input data in a finite element analysis. This is even more so in an AAR study as one needs mechanical, thermal and AAR properties.

This chapter has relied extensively, if not primarily, on a very comprehensive report from the US Bureau of Reclamation in which cores were extracted from AAR affected dams, and properties contrasted with their original values. Henceforth, it should provide the reader with very credible guidance in data preparation.

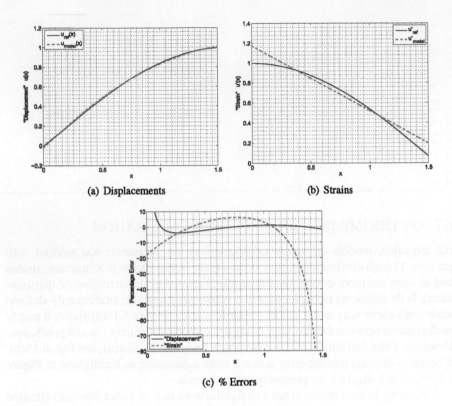

(a) Displacements   (b) Strains

(c) % Errors

FIGURE 6.15 Displacement vs. strain-based model calibration

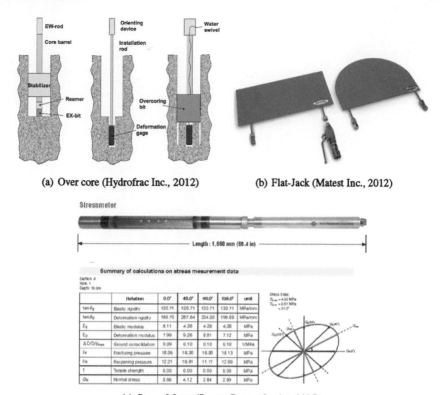

(a) Over core (Hydrofrac Inc., 2012)   (b) Flat-Jack (Matest Inc., 2012)

(c) Stress-Meter (Serata Geomechanics, 2005)

FIGURE 6.16   Field stress measurements

7 Applications

This chapter will explore applications of the AAR constitutive model presented in Chapter 2, as validated in Chapter 4 and enhanced by the parametric study discussed in Chapter 5. It will mainly focus on dams, in introducing nonlinearity through interface elements (see Sect. 3.4), as well as on a massive reinforced concrete structure, in modeling nonlinearity using the smeared crack model (Sect. 3.3).

The first example, an arch gravity dam, will be presented in great detail, given that a comprehensive study has also been performed on this topic. The buttress dam analysis was included in an ICOLD benchmark study, whose limited scope was still able to expose some key considerations. The arch dam is a special case since it is the only one yet to be analyzed either by the author or by the MERLIN computer model he designed (see Appendix B). Instead, it was analyzed by Prof. Mirzabozorg, who implemented the author's published constitutive model (Saouma and Perotti, 2006) in finite element code and performed the analysis using an arch dam. The last example will highlight a massive reinforced concrete structure, considered too expensive to replace, while undergoing AAR.

7.1 ARCH GRAVITY DAM; ISOLA

Analysis performed by L. Perotti and V. Saouma

This section reports on a highly detailed analysis of an arch gravity dam (see Fig. 7.1) and is intended to serve as a reference for the steps required to perform a diagnosis of major infrastructure adversely affected by AAR. The data preparation for such an analysis may prove to be complex and will be reported first, followed by a discussion of results.

7.1.1 DATA PREPARATION

The comprehensive incremental AAR analysis of a concrete dam is relatively complex, regardless of the selected AAR model, given that data preparation for the load can be as cumbersome as it is voluminous (see Fig. 7.2(a)).

The seasonal pool elevation variation (for both the thermal and stress analyses) must first be identified (Fig. 7.2(b)); moreover, the stress-free temperature, T_{ref} (typically either the grouting temperature or the average annual temperature) needs to be identified, along with the external temperature (Figs. 7.2(d)). The pool elevation will affect the internal state of stress, which in turn will alter AAR expansion (through Eq. 2.11, 2.13 and 2.26). This situation is more relevant for high Alpine dams (where the annual pool variation is greater than for low-head, low-altitude dams). This variation will then be replicated over n years for the duration of the analysis (Fig. 7.2(c)). External air and water temperatures will be considered next (Fig. 7.2(d)). In the absence

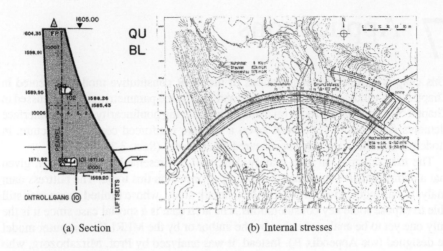

(a) Section   (b) Internal stresses

FIGURE 7.1   Arch gravity dam geometry

of precise field data, the air temperature may be obtained from National Oceanic and
Atmospheric Administration (2013).

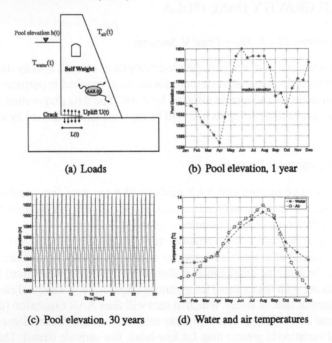

(a) Loads   (b) Pool elevation, 1 year

(c) Pool elevation, 30 years   (d) Water and air temperatures

FIGURE 7.2   Preliminary load data to be collected for the AAR analysis of a dam

The units adopted are MN, Gg and ATU for F, M and T respectively. For time, the ATU (Analysis Time Unit) has been adopted, corresponding to 4.35 weeks or 30.42 days (i.e. 12 ATU span one year).

The next step calls for conducting a transient thermal analysis since the reaction is thermodynamically activated. Consequently, the total temperature is included as part of the constitutive model. Heat transfer by both conduction and convection are taken into account, whereas radiation is implicitly incorporated (Fig. 7.3(a)). The thermal

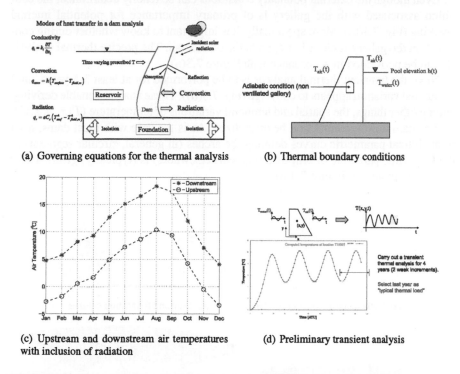

(a) Governing equations for the thermal analysis

(b) Thermal boundary conditions

(c) Upstream and downstream air temperatures with inclusion of radiation

(d) Preliminary transient analysis

FIGURE 7.3   Data preparation for thermal analysis of a dam subjected to AAR

analysis is conducted using the parameters listed in Table 7.1.

TABLE 7.1

Thermal properties found in an arch gravity dam

Reference temperature	T_{ref}	°C	7.0
Conductivity	k	MN.m/(m.ATU.°C)	7.738
Specific heat	c	MN.m/(Gg,°C)	887
Density	r	Gg/m^3	2.510^{-3}

Radiation (Eq. 1.40) is implicitly included by means of a simplified procedure, whereby ambient air temperature is modified (Malla and Wieland, 1999):

$$T_{us} = 0.905T_{air} - 0.4°C \quad \text{Upstream}$$
$$T_{us} = 0.937T_{air} + 7.2°C \quad \text{Downstream} \tag{7.1}$$

resulting in the temperature distribution shown in Fig. 7.3(c).

Even though the external boundary conditions can be readily determined, the condition associated with the gallery is of primary importance for potential internal cracking (Fig. 7.3(b)). More specifically, it is important to know whether during construction the gallery is closed or open to the outside air. The precise thermal analysis should be performed in accordance with Figure 7.3(a).

Next, the transient thermal analysis is to be performed for at least 3-5 years, until the annual variation appears to converge (Fig. 7.3(d)). These analyses enable deriving, among other things, the spatial and temporal variations of temperature ($T(x, y, z, t)$).

The dam however must first be discretized. As is the case with most dams, a set of analytical parametric curves defining the arches (in general, circular segments in the US, while parabolic or elliptical segments elsewhere) is (typically) given. These curves are plotted in Figure 7.4(a).

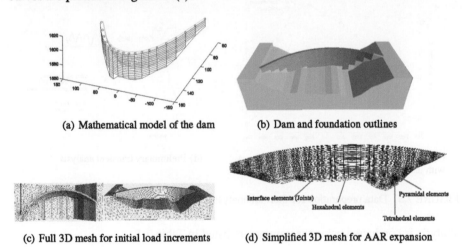

(a) Mathematical model of the dam   (b) Dam and foundation outlines

(c) Full 3D mesh for initial load increments   (d) Simplified 3D mesh for AAR expansion

FIGURE 7.4   3D Finite element mesh for the stress analysis

Using a special program (e.g. Beaver developed by the author), the surface can be discretized. The boundary representation of the dam is then entered into a preprocessor, such as Saouma (2009), in order to generate the finite element mesh. Figure 7.4(c) shows the resulting full 3D mesh used for the stress analysis. Joint elements are placed at both the joints and the rock-concrete interface (Fig. 7.4(d)). Let's point out that a different mesh was required for the thermal analysis, since the interface elements needed to be removed.

As previously indicated, the time increment of one ATU corresponds to 4.35 weeks, and the initial reference temperature was set to zero. Given the external air temperature, both the pool elevation and water temperature boundary conditions were input into this initial boundary value problem. The analysis was performed using MERLIN (B), making it possible to examine the temperature fields. It was found that after four years, the temperature field was harmonic with a one-year frequency. At this point, the analysis was interrupted and $T_{thermal}(x, y, z, t)$ recorded. The computed temperature distribution is shown in Figure 7.5. Note that these temperatures are to be used to evaluate the thermal strains, given that for the AAR analysis the total (i.e. absolute) temperature is needed (as simply obtained by adding the reference temperature of $7°C$).

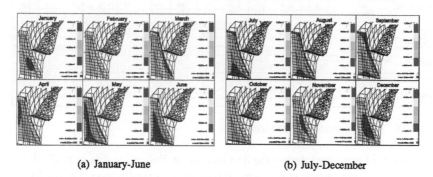

(a) January-June   (b) July-December

FIGURE 7.5   Computed internal temperature distribution variation

Subsequent to the thermal analysis, $T_{thermal}(x, y, t)$ must be transferred to $T_{stress}x, y, t$ since, in general, the same finite element mesh is not available (the foundations, joints and cracks are not typically modeled as part of the thermal analysis).

Lastly, a comprehensive input data file must be prepared for the stress analysis; this file includes:

1. Gravity load (first increment only).
2. $\Delta \dot{T}(x, y, t) = \dot{T}_{stress}(x, y, t) - T_{ref}$ in an incremental format. This is a subtle step that must not be overlooked. The stress analysis is in fact based on the difference between the actual and stress-free temperatures. In addition, an incremental analysis requires this set of data to be provided in an incremental format.
3. Stress-free referenced temperature, which is to be added to the temperature data in order to determine the total absolute temperature needed for AAR.
4. Cantilever and dam/foundation joint characteristics. The former must be included in any arch dam, since expansion may lead to an upstream joint opening, while the latter must be taken into account given that AAR-induced swelling may result in a separation of the dam from the foundation in the central portion of the foundation.

5. Uplift load characteristics (typically matching the upstream hydrostatic load).

6. AAR data, which has been described above.

Moreover, the compiled set of data must be looped over at least 50 years in order to provide a complete and correct set of natural and essential boundary conditions (Fig. 7.6).

		January		February		March		April		May		June	
Incr.		6.00	7.00	7.00	8.00	9.00	10.00	11.00	12.00	13.00	14.00	15.00	16.00
Body force		dam											
Hydrostatic	Pool Elevation	1596.47	1593.53	1593.53	1592.94	1590.59	1589.71	1588.24	1586.76	1591.47	1598.24	1602.65	1604.00
	Incremental Elevation	-5.03	-2.94	0.00	-0.59	-2.35	-0.88	-1.47	-1.47	4.71	6.76	4.41	1.35
Uplift	Pool Elevation	1596.47	1593.53	1593.53	1592.94	1590.59	1589.71	1588.24	1586.76	1591.47	1598.24	1602.65	1604.00
	Incremental Elevation	-5.03	-2.94	0.00	-0.59	-2.35	-0.88	-1.47	-1.47	4.71	6.76	4.41	1.35
Temperature [°C]	Air	-3.10	-2.14	-1.67	-1.43	0.24	1.90	2.14	2.38	4.76	6.90	8.10	8.81
	Water	1.00	1.00	1.00	1.00	1.00	1.50	3.00	3.00	5.00	6.00	8.00	8.00
		July		August		September		October		November		December	
Incr.		17.00	18.00	19.00	20.00	21.00	22.00	23.00	24.00	25.00	26.00	27.00	28.00
Body force		dam											
Hydrostatic	Pool Elevation	1602.35	1602.65	1602.65	1602.65	1600.59	1595.29	1595.88	1593.24	1596.76	1598.53	1598.24	1601.50
	Incremental Elevation	-1.65	0.29	0.00	0.00	-2.06	-5.29	0.59	-2.65	3.53	1.76	-0.29	3.26
Uplift	Pool Elevation	1602.35	1602.65	1602.65	1602.65	1600.59	1595.29	1595.88	1593.24	1596.76	1598.53	1598.24	1601.50
	Incremental Elevation	-1.65	0.29	0.00	0.00	-2.06	-5.29	0.59	-2.65	3.53	1.76	-0.29	3.26
Temperature [oC]	Air	9.76	10.24	11.43	12.38	11.43	10.24	6.67	3.57	0.95	-1.19	-2.62	-4.05
	Water	9.00	10.00	11.00	11.00	11.00	8.50	6.00	4.00	3.00	3.00	2.00	1.00
AAR	AAR	Activated											

FIGURE 7.6   Data preparation, cyclic load

The dead load is applied during the first increment. Following this step, displacements are reset to zero, while maintaining the internal strains/stresses. During increments two through five, the hydrostatic (and uplift) load is applied, and the AAR expansion only initiates at increment six.

7.1.2 STRESS ANALYSIS

Following completion of the transient thermal analysis, the stress analysis may be performed. It should be noted however that the finite element mesh for the stress analysis of a dam affected by AAR must differ from the mesh used for the thermal analysis and moreover includes joints, the interface between dam and rock foundation, and the rock foundation. These components are not required in the thermal analysis but are very important to capturing the real behavior of a dam affected by AAR (and thus capturing the real crest displacements on which parameter identification is based, as will be explained in the following section). AAR expansion can indeed result in: 1) opening of the downstream vertical joints and closure of the upstream vertical joints in an arch dam; 2) possible movement of the various buttresses on a gravity dam along the joints; and 3) sliding of the dam when subjected to a compressive state of stress on the foundation joint.

With regard to the temporal and spatial variations of temperature, it should be kept in mind that the stress analysis requires a temperature difference with respect to the stress-free temperature (namely the grouting temperature $T(x, y, z) - T_{grout}$), whereas AAR evolution depends on the total absolute temperature inside the dam $T(x, y, z)$.

7.1.3 RESULTS

Since the "proper" AAR parameters were not known *a priori*, the system identification process described in Section 6.6 was adopted with the initial and final values shown in Table 7.2. Figure 7.7 illustrates the real-time graphical display of the evolution

TABLE 7.2

Initial and final parameter values in system identification study; Reference temperature 60°C

Parameter	Initial	Final	Unit
$\varepsilon^\infty \times 10^3$	3.8947	3.5500	-
τ_C	1.564	1.453	ATU
τ_L	3.279	4.072	ATU

of the three key parameters as well as the comparison between computed and actual crest displacements. Each data point, constituting a full 30-year analysis (or about 360 increments), is thus a CPU-intensive operation (requiring a few days of CPU time).

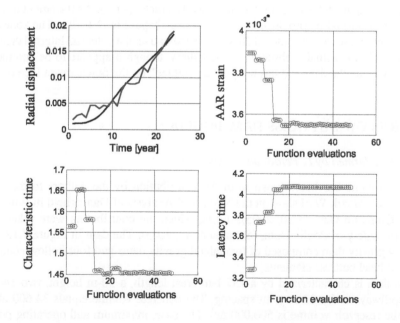

FIGURE 7.7 System identification process for the AAR analysis of an arch gravity dam

The final results are shown in Figure 7.8(a), where the field measurement of crest displacements are plotted along with the first and last numerical simulations. These results are "sharper" upon examination of Figure 7.8(c). In either case, given the complexity of the simulation and the randomness of actual material properties (as "smoothed" through the system identification process), the results can be qualified as "good" and then used for prediction purposes.

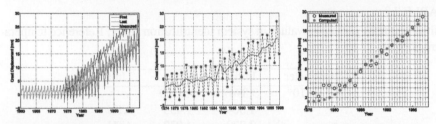

(a) System identification and re-   (b) Field-recorded crest displace-   (c) Final annual comparison
sults with seasonal variation   ments

FIGURE 7.8   Crest displacement comparison

The stress and strain comparisons are shown in Figure 7.9. A zone of high tensile stresses can be noted at the center of Block 10.

Of special interest herein are the stresses and displacements along the base, as indicated in Figure 7.10(a). Let's note the positive crack opening displacement of the joint, with confirmation provided by the deformed shape (Fig. 7.10(b)). The zone of high tensile stresses at the center of Block 10 is indicative of a potential internal crack. Such a crack was actually observed in the gallery, though it appears to be confined (for now) inside the dam and has not yet "daylighted" on the downstream face (Fig. 7.11).

7.2   HOLLOW BUTTRESS DAM; POGLIA

Analysis performed by L. Perotti and V. Saouma

This example has been drawn from the contribution by the authors to the 8th ICOLD Benchmark Workshop on the Numerical Analysis of Dams (held in Wuhan, Hubei, China on Oct. 23-30, 2005), during which the contributors were asked to assess the safety of Poglia Dam, (Saouma et al., 2005). The dam in Figure 7.12 is a hollow gravity dam composed of two lateral gravity dams and four hollow central diamond-head buttress elements.

This dam is characterized by a 137.1-m crest length, a 50-m height, two 14-m wide spillways and a 22-m joint spacing. The concrete volume equals 34,600 m^3, while the reservoir volume is 500,000 m^3. The base, maximum and operating pool elevations are 628.1, 632 and 625-628 m, respectively. The concrete features: a mass density of 2,400 kg/m^3, an elastic modulus of 18,000 MPa, a Poisson's ratio of 0.2, and compressive and tensile strengths of 32 and 1.5 MPa respectively. Accelerated

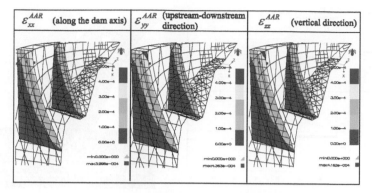

(a) Internal strains caused by AAR

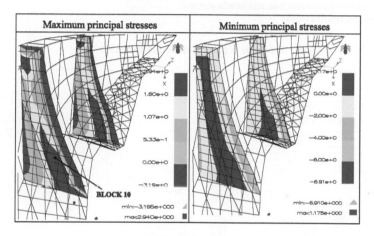

(b) Internal stresses

FIGURE 7.9   Internal AAR and stresses

expansion tests were conducted on cores extracted from Poglia Dam at both 38 and
80°C. From these tests, it was determined that the residual expansion equals approx.
0.095%. Rock has an elastic modulus of 10,000 MPa and a Poisson's ratio equal to
0.1. The compressive and tensile strengths amount to 32 and 0.15 MPa, respectively.
Let's also closely observe the rock/concrete joint, with a friction angle of 37 degrees
and a cohesion of 1.0 MPa.

The dam monitoring system is essentially based on: 1) 12 monitoring points along
the dam for recording vertical displacements, 2) 6 targets for recording horizontal crest
displacements in the stream direction, and 3) 2 pendulums for recording both longitudinal and lateral displacements. The targets have been placed on either side of the
vertical joints. Additional instrumentation consists of extensometers and piezometers.

Though no chart exists to plot crest displacement vs. time, it is understood that
the crest vertical drift (which started in 1970) reached 8 mm in 1982 and then a total

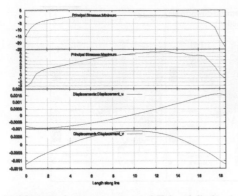
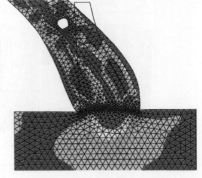

(a) Minimum/maximum stresses (MPa) and displacements (m) along the rigid rock/concrete interface; maximum principal stresses (MPa)

(b) Deformed shape of dam section

FIGURE 7.10   "Lift-off" of central portion of dam due to AAR

(a) Internal crack

(b) Internal crack

FIGURE 7.11   Observed internal crack

value of 30 mm in 2000. This increase is quasi-linear with a slightly greater rate
during the later years (Fig. 7.13). In addition, irreversible displacements of about 0.1
mm/year and 0.2 mm/year were recorded in the cross-valley direction (i.e. toward the
right side) and along the stream direction, respectively. A downstream displacement
recorded on the right-hand corner may prove to be the most worrisome finding (due
to lateral expansion of the two adjacent blocks).

7.2.1   TRANSIENT THERMAL ANALYSIS

Since our AAR model is highly dependent on temperature, a preliminary transient
thermal analysis needs to be performed. In this analysis, only the dam is considered.

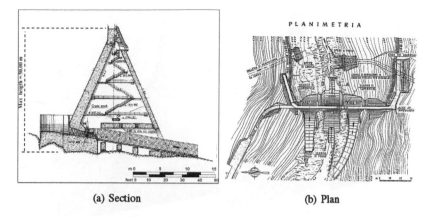

(a) Section   (b) Plan

FIGURE 7.12   Section (central element) and layout of Poglia Dam

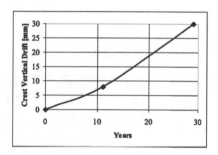

FIGURE 7.13   Measured crest vertical displacement

Half the central element was modeled, and due to symmetry, adiabatic boundary conditions could be applied on the plane of symmetry. Convection was neglected, and the recorded temperatures were assigned on the faces exposed to air (both outside and inside) and to water (Fig. 7.14). Conductivity k was set to 44.67 MN.m/(mATU.°C) and the mass density to $2.43 10^{-3}$ Gg/m^3. In the context of our analysis, one ATU (Analysis Time Unit) corresponded to 6 months, hence 2 time increments per year were used. The temperatures are listed in Table 7.3. It should be noted that the average (stress-free condition) temperature was subtracted from these values when conducting the thermal analysis. This average temperature was subsequently added to the nodal temperature for the AAR analysis (given that expansion depends on the total temperature and not on the temperature differential). The transient analysis was performed for five years; since results stabilized after four years, the Year 5 results were considered to be representative of the annual temperature variation. Figure 7.15 displays the internal dam temperature.

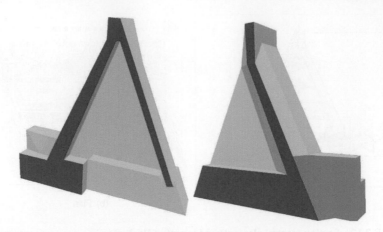

FIGURE 7.14   Adiabatic boundary conditions for the thermal analysis

TABLE 7.3
Temperatures used for the Poglia Dam analysis

Season	Inside	Outside	Water
Winter	4.4	0.1	2.9
Summer	19.2	29	12.6

7.2.2   STRESS ANALYSIS

Following the thermal analysis, a stress analysis was conducted using a mesh that included the dam, its foundation and the rock/concrete interface joints.

The material properties adopted for the analysis are those specified by the Benchmark, with one notable exception. The tensile strength and cohesion of the dam/foundation interface were increased to 1.5 and 1.0 MPa (rather than the specified 0.15 and 0.10 MPa) because the preliminary analysis had indicated that these specified values were too small to resist the pool elevation. It should be noted that only initial guesses were introduced for the maximum volumetric strain latency and characteristic time, since these values were to be determined subsequently from an automated system identification procedure (see Section 6.6). Following convergence of this minimization (of the error between computed and measured crest displacements), "fine-tuning" was manually performed to further reduce this error. The final (numerically obtained) volumetric strain was 3.50×10^{-3}, whereas the experimentally obtained value (from laboratory tests) was 0.95×10^{-3}. Total dam expansion however could only be justifiably determined from the numerical calculation.

The gravity load (for the dam only) was applied in the first increment and the displacements (but not the stresses) were subsequently reset to zero (to remove

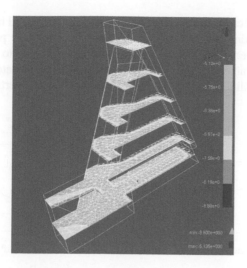

FIGURE 7.15   Internal temperature of Poglia Dam

self-weight displacements from the results). The pool elevation was then raised from base elevation 582.10 m to maximum elevation 628.09 m in four increments of 11.5 m. each. It should be noted that as the pool elevation was increased, not only did the hydrostatic pressure rise, but so did the uplift pressure distribution under the dam. This pressure distribution will be automatically adjusted from triangular to trapezoidal should the dam base crack. From the sixth increment forward, pool elevation was kept constant and the AAR was active (increments 6 and 7 correspond to winter and summer of the first year). An analysis corresponding to 55 years of dam operations was carried out and the resulting crest displacements were compared with measurements from the inverse analysis.

7.2.3   ANALYSIS AND RESULTS

The 3D finite element mesh contained 6,145 nodes and 24,133 elements. The 3D nonlinear analysis employed the Tangent Stiffness method along with a Line Search and comprised 115 increments. The convergence criteria were set to 0.01, 0.02, 0.05 and 0.02 for the Energy, Relative Residual (ratio of the Euclidian norm of the current to the initial residual load vectors), Absolute Residual (ratio of the infinity norm of the current to the initial residual load vectors), and Displacement Error, respectively. A 50-iteration maximum was allowed. Each analysis lasted roughly 7 hours on a 3.00-GHz Xeon processor with 1.5 GB of RAM.

Figure 7.16(a) shows the crest displacements for the first seven increments, thus highlighting the crest displacement resulting from gravity and resetting the displacements to zero, followed by the four increments associated with impounding. Crest displacements over the entire analysis are provided in Figure 7.16(b); let's note that while the measured crest displacement has not been captured, all later displacements have been (accounting for about 3 cm). This situation may be partially due to the scarcity of field observations (moreover, indication is lacking as to whether they were taken during the summer or winter) as well as to the fact that the lateral and significant dam displacement/sliding has been neglected. shows both the vertical and horizontal

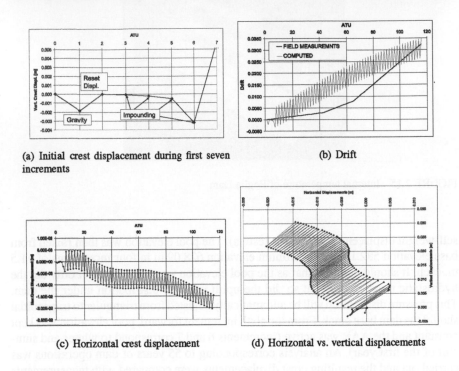

(a) Initial crest displacement during first seven increments

(b) Drift

(c) Horizontal crest displacement

(d) Horizontal vs. vertical displacements

FIGURE 7.16   Analysis results for the Poglia Dam

crest displacements. Let's note that the seasonal variations are not negligible (possibly due to the thermal gradient between air trapped inside the dam and outside air), as compared to the total irreversible AAR variation. As expected (due to lateral stresses), the horizontal displacements are substantially smaller than vertical displacements.

The overall deformed shape however is somewhat puzzling (Fig. 7.17). It appears that for the given numerical assumptions, most of which have been stipulated by the Benchmark organizers, the dam is failing because of sliding. The sharp curvature of the top deformation corresponds to the pool elevation, which indeed lies 2 m below the crest.

Figure 7.18 illustrates various contour lines for the main rock/concrete joint. It is interesting to note that the crack opening displacements are maximum in the middle and then minimum on the upstream/downstream side. This finding had already been observed in the arch gravity dam analyzed previously. The normal stresses are mostly zero, and full uplift develops over most of the joint. The joint sliding displacement

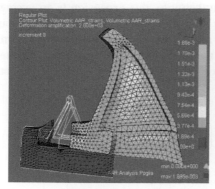

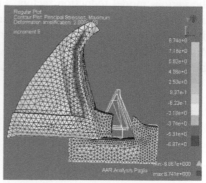

(a) Volumetric AAR strain   (b) Maximum principal stresses (MPA)

FIGURE 7.17   Deformed shape of the Poglia Dam

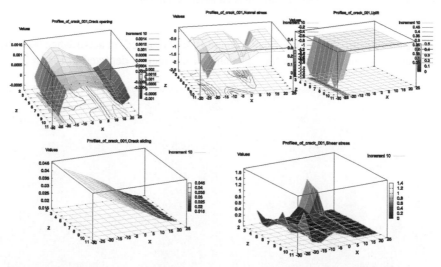

FIGURE 7.18   Contour plots of the main rock/concrete joint, joint opening displacements (m), normal stresses (MPa), uplift pressures (MPa), joint sliding displacements (m) and joint shear stresses (MPa)

is linear and equal at most to 4.5 cm (note that adopting the Line Search method may have constrained overall sliding to such a "small" value). The shear distribution over the joint can also be visualized in this figure. According to these indications, the global safety factor relative to sliding remains below 1.

The overall AAR expansion at the end of the analysis ε^{∞} was 1.65×10^{-3}, whereas the actual AAR expansion inside the dam was about 4×10^{-4}; this difference is due to the various model reduction factors. Furthermore, this expansion is on average equivalent to 40°C, i.e. just slightly lower than the 42°C determined in a previous study, which did not model the interface or assume thermal expansion to be uniform.

A second analysis was conducted using the lateral faces of the dam unconstrained in the lateral (Z) direction. Figure 7.19 shows that with the lateral faces of the dam

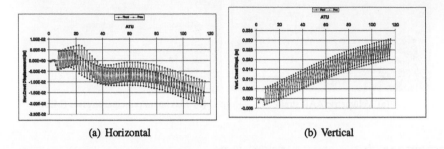

(a) Horizontal   (b) Vertical

FIGURE 7.19   Effects of lateral constraints on AAR expansion

now free to expand, the vertical displacement is now reduced, in accordance with the premises of our model. The deformed shape of the dam is depicted in Figure 7.20,

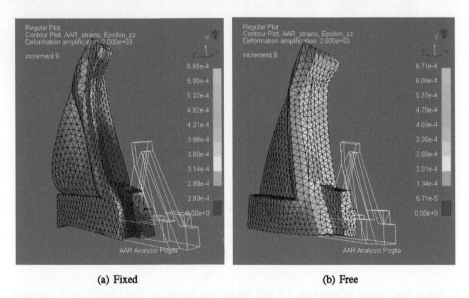

(a) Fixed   (b) Free

FIGURE 7.20   Deformed mesh and AAR strains in the lateral directions for both free and fixed boundary conditions

where the lateral swelling is quite obvious. Despite the extensive effort for this analysis, the authors are not convinced that the structural model adopted actually reflects the current state of the dam. According to the numerical model, the dam has at the very least partially failed (though we know it is still "alive and well"). This perspective has rendered any subsequent evaluation of the failure pool elevation redundant, especially given the model's limited scope. Let's emphasize that the concern here is not the selected AAR constitutive model, but rather some of the physical assumptions, including:

- Rock/concrete interface properties. The cohesion was specified to be 0.15 MPa by the Benchmark organizers, which is a very low value. Moreover, preliminary analysis has indicated that the dam could not even sustain the hydrostatic load, let alone AAR expansion. The cohesion and tensile strength values were thus arbitrarily increased by a factor of 10.
- The"2D" plane strain assumption may not actually reflect the complex stress state in this oddly shaped dam.
- This dam appears to be overly sensitive to thermal loads, hence an accurate long-term prediction of the dam would have required smaller time steps than what was adopted (6 months due to computational and "human" constraints).
- The benefit of a complete model is further justified by the most likely failure mechanism shown.

In light of the above assumptions and for a more accurate long-term prediction of this dam, the author would suggest that any future analysis considers the following:

1. Modeling of the entire dam, since opening of the joint connecting the dam to the lateral gravity dams may be the primary concern (Fig. 7.21), which in turn could cause overall structural failure.
2. Smaller time steps for the thermal and AAR analyses, accompanied by a better grasp of the temperature boundary conditions.
3. Improved investigation of the rock/concrete interfaces through core extractions and testing, in addition to piezometer readings of the uplift pressures.
4. Possibly, but not necessarily, accelerated tests of concrete cores at either 38°C or 70°C.
5. Again possibly, but not necessarily, "calibration" of the dam response by carefully monitoring crest displacements and the internal stress state using flat jacks (an appropriate step given the relatively small thickness of the structural components) during dam drainage and impounding.

Only with these considerations can the failure pool elevation be realistically determined, as any competing approach would serve as a mere academic exercise useful for comparing models but not for performing an accurate evaluation. Conducting a full 3D analysis of this dam would simply offer speculation on the likely mode of failure mechanism. The sharp corner between the two dam segments constitutes, in this case, the dam's "Achilles' heel", a worrisome feature that warrants further investigation.

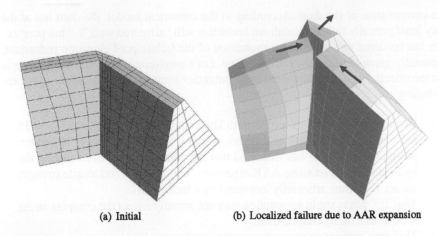

(a) Initial   (b) Localized failure due to AAR expansion

FIGURE 7.21   Possible failure mechanism at a joint connecting two parts of the dam

7.3   ARCH DAM, AMIR-KABIR

Analysis performed by Prof. H. Mirzabozorg; reproduced here with permission

The author's constitutive model, (Saouma and Perotti, 2006), was implemented by Mirzabozorg (2013) in his nonlinear finite element code (NSAD-DRI) and subsequently used to diagnose the Amir Kabir Dam in response to a *hypothetical* AAR expansion.

7.3.1   DAM DESCRIPTION

Amir Kabir is an arch dam (Fig. 7.22(a)) some 180 m high, with a crest length of 90 m and a crest width of about 7.85 m; it was built between 1957 and 1961. The 750,000 m³ of concrete used aggregates supplied from nearby rock deposits. It is *assumed* that the dam undergoes an AAR reaction, and its eventual response will be compared with the actual response based on the same computer model (in excluding AAR).

In both models, the self-weight, hydrostatic pressure on the upstream face and temperature loads are all identical. Due to the small dam-foundation interface thickness, uplift was neglected. In the analysis, the pool elevation fluctuated in a harmonic fashion between the maximum and minimum elevations (164 m and 136 m, respectively). The finite element mesh was composed of 72 three-dimensional (20 nodes, 27 Gauss points) cubic elements and a 592-node network (only one element through the dam thickness); the galleries and other structural details were excluded.

At every time step, the internal temperature was computed by finite differences using: $k = 2.4 \text{ m}^2/\text{month}$ and $\alpha = 8.05 \times 10^{-6}/^{\circ}\text{C}$. The reference temperature (corresponding to the stress-free temperature) is shown in Figure 7.23, (Akhavan, 2001).

(a) Downstream view of the Amir Kabir Dam

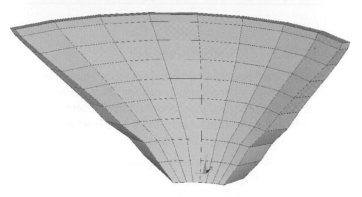

(b) FE model of the dam body

FIGURE 7.22   Dam and mesh description

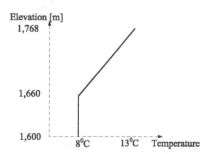

FIGURE 7.23   Reference temperature distribution

The AAR properties are listed in Table 7.4.

TABLE 7.4

AAR model parameter values

Parameter	value	parameter	value
U_c	5,400	U_L	9,400
β_E	0.8	β_f	0.8
γ_t	0.18	γ_t	0.1
β	0.5	ε	0.00284
α	1.3333	θ_0	333
τ_{c0}	2.7	τ_{l0}	5.5
σ_u	10^6	f_c	30.5×10^6

7.3.2 ANALYSIS RESULTS AND DISCUSSION

The analysis was performed for a 25-year period with a one-month increment (i.e. 300 total time steps). The results are presented as contour lines in Figures 7.24, 7.25 and 7.26, which reveal the minimum and maximum principal strains for both analyses on the upstream, middle and downstream faces. Let's note how AAR has dramatically increased the strains and altered their distribution.

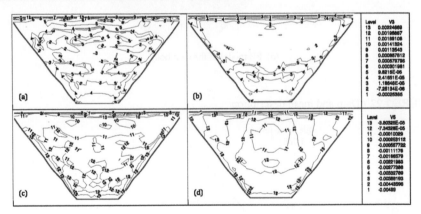

FIGURE 7.24   Contours of principal strains on the upstream face: a) minimum and b) maximum without AAR; c) minimum and d) maximum with AAR

Figure 7.27 shows the maximum compressive and tensile strains on the upstream, middle and downstream faces both with and without AAR. Once again, we can observe the drastic increase in strains due to AAR.

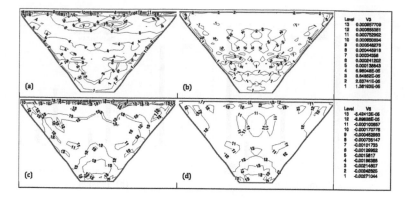

FIGURE 7.25   Contours of principal strains on the middle face: a) minimum and b) maximum without AAR; c) minimum and d) maximum with AAR

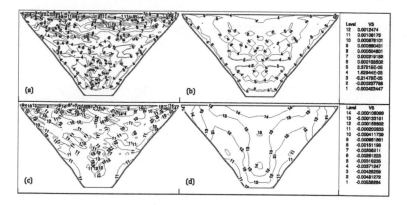

FIGURE 7.26   Contours of principal strains on the downstream face: a) minimum and b) maximum without AAR; c) minimum and d) maximum with AAR

The high tensile strains reported above wind up inducing internal cracks, as highlighted during the nonlinear analysis described in Figure 7.28, which indicates cracked Gauss points on the upstream, middle and downstream faces of the dam. Four major cracks are observed to cross 25 elements. The first crack is detected at the 263rd time step (21st year). Let's also note that the analysis without AAR did not yield any cracking.

Moreover, the temporal variation in crest displacement provides the best indication of AAR occurrence. Figures 7.29(a) and 7.29(b), i.e. the vertical and horizontal crest displacements for both analyses, are quite contrasted. Not until the 230th increment (19th year) do AAR results show a substantial increase in displacement.

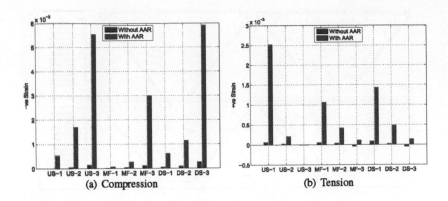

FIGURE 7.27   Maximum compressive and tensile strains on the upstream, middle and down-stream faces)

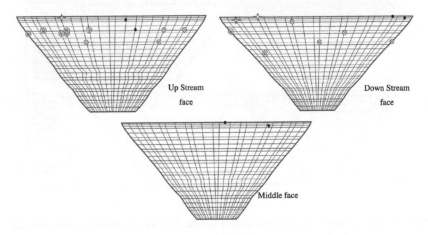

FIGURE 7.28   Location of cracks at Gauss points on the upstream, middle and downstream faces

7.4   ARCH DAM, KARIBA

Analysis performed by Prof. Pan, Tsinghua University, Beijing; reproduced herein with permission

It was mentioned above that one of the main advantages of the AAR constitutive model presented in Chapter 2 is its capacity to bind with other concrete/material constitutive models. This is in fact the approach adopted by Pan et al. (2013), who followed this protocol in analyzing the Kariba Dam (whereby the author's AAR model was combined with a plastic-damage model).

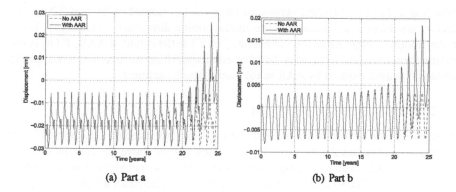

(a) Part a   (b) Part b

FIGURE 7.29 Monthly variation in horizontal crest displacement both with and without the AAR effect

7.4.1 CONCRETE CONSTITUTIVE MODEL

According to the elasto-plastic model, total strain ε is defined as the sum of elastic strain ε^e, plastic strain ε^{pl}, thermal strain ε^{th}, creep strain ε^{ve} and the strain due to AAR ε^{aar}:

$$\varepsilon = \varepsilon^e + \varepsilon^{pl} + \varepsilon^{th} + \varepsilon^{ve} + \varepsilon^{aar} \tag{7.2}$$

Each term will now be briefly described separately.

The plastic strain rate is determined from the classical non-associated potential flow rule:

$$\dot{\varepsilon}^{pl} = \dot{\lambda} \frac{\partial Q(\bar{\sigma})}{\partial \bar{\sigma}} \tag{7.3}$$

where λ is the plastic multiplier, and $Q(\bar{\sigma})$ the non-associated flow potential.

The thermal strain is simply defined by:

$$\varepsilon^{th} = \alpha (T - T_{ini}) \delta \tag{7.4}$$

where α is the coefficient of thermal expansion, T_{ini} the initial temperature, and δ Kronecker's delta.

A three-dimensional version of the Kelvin-Voigt model was introduced to consider the effects of long-term creep:

$$\varepsilon^{ve}(t + \Delta t) = e^{-\frac{E_{ve}}{\eta_{ve}} \Delta t} . \varepsilon^{ve}(t) + \frac{[A]\,\bar{\sigma}(t)}{E_{ve}} \left(1 - e^{-\frac{E_{ve}}{\eta_{ve}} \Delta t} \right) \tag{7.5}$$

where $\bar{\sigma}$ is the effective stress, E_{ve} the rheological model modulus, η_{ve} the viscosity, Δt the incremental time step, and $[A]$ the matrix of Poisson's ratio coefficients.

Lastly, AAR strain is defined as:

$$\varepsilon_i^{aar}(t) = W_i(\sigma_i, \sigma_j, \sigma_k) . \xi(t, T) F(h) \varepsilon_v^{\infty} \tag{7.6}$$

which is a slightly simplified version of the model presented in Section 2.2. σ_i, σ_j and σ_k are the principal stresses in the corresponding directions i, j and k, W_i is the redistributing weight along the i direction, $\xi(t, T)$ is the extent of AAR, $F(h)$ is the relative humidity, and ε_v^∞ is the material constant specifying the maximum volumetric expansion strain induced by AAR in the stress-free experiment.

7.4.2 DESCRIPTION OF THE DAM

Kariba Dam, a concrete arch dam completed in 1959, is located on the Zambezi River between Zambia and Zimbabwe. The dam rises 128 m above the foundation. The structure began to exhibit swelling soon after the startup of operations. The monitoring system was installed based on a topographic survey conducted for a number of targets placed on the downstream face. The main instrument locations are shown in Figure 7.30. The actual measured dam displacements during the period 1963–1995 have been used for model calibration and the subsequent response has been predicted numerically. The dam-foundation system was discretized with 20-node brick elements

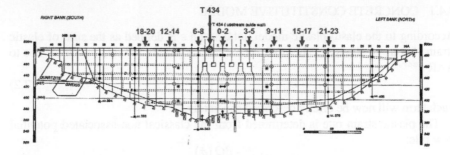

FIGURE 7.30   Locations of motoring points on the dam crest

and 15-node prism elements (Fig. 7.31).

The mechanical properties of the concrete dam are as follows: $E_0 = 22GPa$, $\nu = 0.2$, $f_t = 2.1MPa$, $f_c = 25.0MPa$, $G_f = 200N/m$, $\rho = 2350kg/m^3$. The foundation rock properties are: $E_0 = 10GPa$, $\nu = 0.2$. The foundation rock is assumed to be linear elastic. Moreover, the AAR kinetics parameters are: $\tau_L = 25$ years, $\tau_c = 25$ years, and $\varepsilon_v^\infty = 0.0028$.

7.4.3 ANALYSIS

The self-weight of this dam is initially applied according to the assumption of independent cantilevers and joint grouting at a uniform temperature equal to the average ambient air temperature. The annual average temperature at the site is $27^\circ C$ with low seasonal variations; consequently, the thermal effects are neglected in the long-term analysis. The water pressure exerted on the upstream face of the dam varies with the reservoir water level during the impounding and operating periods. The reservoir water level variation is indicated in Figure 7.32. The incremental time step is set at

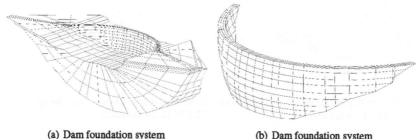

(a) Dam foundation system   (b) Dam foundation system

FIGURE 7.31   FE mesh for the Kariba Dam

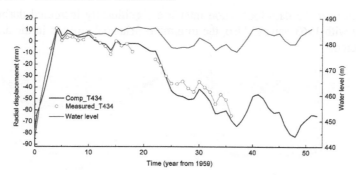

FIGURE 7.32   Radial displacements at the crest of the crown cantilever (T434)

14 days and a 52-year duration covering 1959 to 2012 is analyzed. Figures 7.33(a) and 7.33(b) show the vertical displacements at the various monitoring points. The displacements obtained from the simulation in 1963 have been initialized to zero and the subsequent values have been adjusted so that the computed results can be directly compared with field measurements. Vertical displacements increase gradually during the AAR process.

7.4.4   OBSERVATIONS

We can observe that the expansion deformation is larger over the central portion compared with the dam blocks on both sides. A similar response is derived from the analysis, with computed displacements at the monitoring points (except for 0-2 and 3-5) revealing reasonable accuracy (Fig. 7.33).

The principal stress distributions of the dam in both 1963 and 2012 are displayed in Figures 7.34. The reservoir was first impounded to the operating water level in 1963, and in 2012 the pool elevation had remained the same. A stress redistribution has clearly occurred due to AAR expansion, and the maximum principal stresses in the central portion of the downstream face and in the dam heel were found to be high in 1963, though these stresses had been considerably reduced (from tensile to compressive state) in 2012. The maximum principal stresses on the downstream face

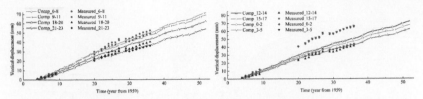

(a) Vertical displacements at monitoring points (b) Vertical displacements at monitoring points
6-8, 9-11, 18-20 and 21-23   0-2, 3-5, 12-14 and 15-17

FIGURE 7.33   Computed response

of the dam near the dam-foundation interface significantly increased during 2012
compared with 1963 values. Also, the minimum principal stresses in the dam have
been lowered.

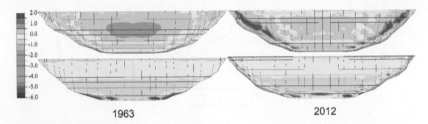

(a) Distribution of the maximum principal stress (MPa) on both the downstream and upstream
faces

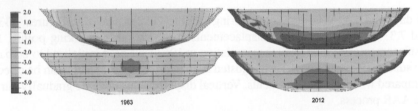

(b) Distribution of the minimum principal stress (MPa) on the downstream and upstream faces

FIGURE 7.34   Principal stresses

Furthermore, the predicted crack pattern is shown in Figure 7.35. At the dam heel,
a stress concentration exists whenever hydrostatic pressure is applied, and localized
cracks have apparently been forming there since 1963. As a result of AAR, indications
suggest that cracking at the dam heel has extended even further, and severe cracking
and damage are predicted to occur on the downstream face near the dam-foundation
interface on both sides in 2012 (following 52 years of operations).

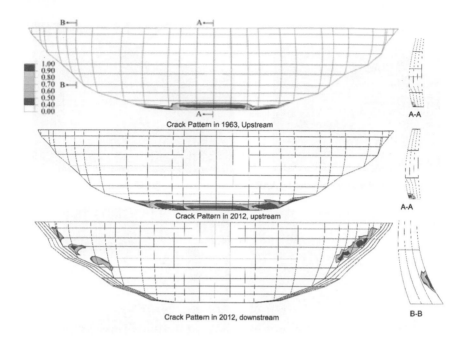

FIGURE 7.35   Predicted crack patterns for the Kariba Dam

7.5 MASSIVE REINFORCED CONCRETE STRUCTURE

Analysis performed by Y. Yagome, H. Noguchi (Tokyo Electric Power Company) with V. Saouma; reproduced here with permission

7.5.1 DESCRIPTION

AAR makes no distinction with respect to the type of structure it will affect; its effects however are most drastic in massive (in terms of volume) concrete structures that are regularly subject to monitoring, e.g. dams. Other unmonitored structures may also be adversely affected by AAR expansion, but only the most strategically important structures receive proper attention from their owners. Such is the case for the massive reinforced concrete support of a high-voltage transmission tower with 1.5 m by 1.5 m cross-girders (13.7 m span) and 2.0 m by 2.0 m columns. The girders are heavily reinforced both longitudinally (14 H 29 and 14 H 22) and vertically (H 16 spaced at 250 mm).

An unusual crack pattern, within a localized section, was observed with AAR expansion being suspected (Fig. 7.36). This pattern was later confirmed through laboratory tests.

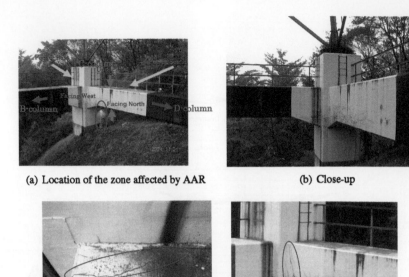

(a) Location of the zone affected by AAR   (b) Close-up

(c) Structural cracks along with FE simulation results

FIGURE 7.36   Heavily reinforced transmission tower foundation partially affected by AAR

Figures 7.37(a) and 7.37(b) show the geometry, loading and dimensions of this massive structure. The corner transmission tower exerts both tensile and compressive loads on its supports.

Of particular concern herein is girder CD, where AAR had been observed at one end. Figure 7.38(a) illustrates the field-recorded cracks on the four faces, while Figure fig:aar-TT-9 shows the AAR strain measurements recorded using a procedure described in Section 1.2.5 (crack index). This massive structure is also heavily reinforced (Fig. 7.39).

7.5.2 MODEL

Four finite element models have been considered:

Model	Model	Extent of Expansion
1	Frame C-D	Entire Girder C-D
2	Entire	Entire Girder C-D
3	Frame C-D	Localized in Girder C-D
4	Entire	Localized in Girder C-D

and the finite element mesh (6,271 elements and 3,176 nodes) associated with the 3D model is shown in Figure 7.40. Let's point out the heavy internal reinforcement that

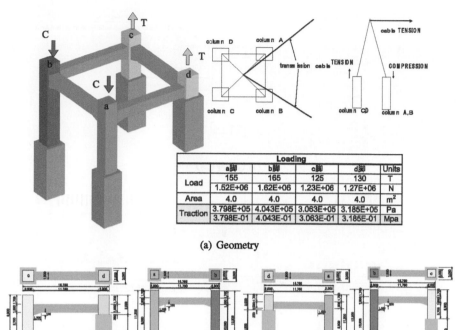

Loading					
	a欄	b欄	c欄	d欄	Units
Load	155	165	125	130	T
	1.52E+06	1.62E+06	1.23E+06	1.27E+06	N
Area	4.0	4.0	4.0	4.0	m²
Traction	3.798E+05	4.043E+05	3.063E+05	3.185E+05	Pa
	3.798E-01	4.043E-01	3.063E-01	3.185E-01	Mpa

(a) Geometry

(b) Dimensions

FIGURE 7.37   Geometry of the massive reinforced concrete structure

has been explicitly modeled along with the Winkler springs surrounding a portion of the columns buried below ground. This becomes an important consideration in the presence of a large lateral load.

7.5.3   RESULTS

The results for each of the four analyses will be presented separately.

For case 1, Fig. 7.41(a): No cracks are observed in girder C-D or columns D. On the other hand, an unreasonably large crack is found in column D.

For case 2, Fig. 7.41(b): A few cracks are observed in C-D, no cracks in column D, and large cracking in column C.

As for case 3, Fig. 7.41(c): Excessive cracking is noted in girder C-D and in column C.

Lastly, for case 4, Fig. 7.41(d): The cracking is finally close to the observed pattern, and only a few cracks are computed in column C.

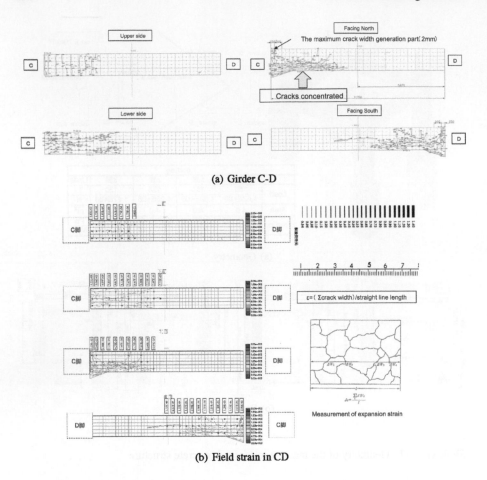

(a) Girder C-D

(b) Field strain in CD

FIGURE 7.38   Field strain measurements

The computed crack opening displacement can now be compared with the measured displacement (through the crack index) in Figure 7.42; this comparison confirms the earlier qualitative assessment that case 4 indeed yields the closest values.

This finding has further been confirmed by a closer examination of the damaged zone in analysis 4, Fig. 7.43, where the strains are in fact very close.

Moreover, the temporal evolution of strains is shown in Figure 7.44.

7.5.4   SEISMIC ANALYSIS FOLLOWING AAR EXPANSION

Following a static analysis of the structure, which has more or less captured both the observed crack pattern and strain recordings in the field, a seismic analysis was performed.

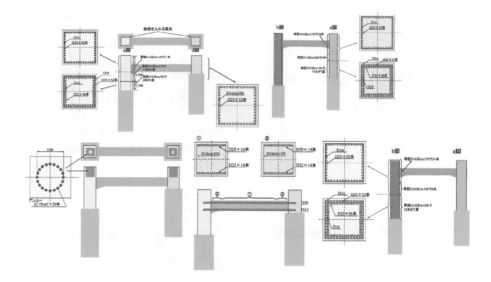

FIGURE 7.39   Reinforcement

FIGURE 7.40   Finite element mesh

It would be erroneous to undertake such an analysis simply with the degraded tensile strength and Young's modulus, since the "locked-in" stresses caused by AAR and existing cracking would not be taken into account. Instead, the analysis was performed via a "restart" feature in the Merlin finite element code. Per owner's request however, these results cannot be published.

7.6   SUMMARY

AAR engineering simulation is of paramount importance to engineers confronted with a degraded structure. The most important questions concern how long the structure

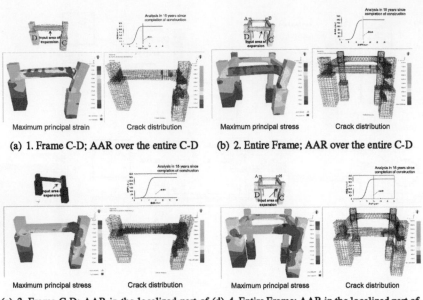

(a) 1. Frame C-D; AAR over the entire C-D   (b) 2. Entire Frame; AAR over the entire C-D

(c) 3. Frame C-D; AAR in the localized part of (d) 4. Entire Frame; AAR in the localized part of
C-D   C-D

FIGURE 7.41   Finite element analysis of various cases

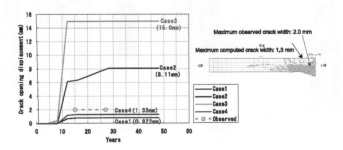

FIGURE 7.42   Finite Element Analysis summary; COD

will continue to expand (given constant exposure to the reactive alkali and aggregates),
and what would be the accompanying structural response. This chapter has presented
a rigorous procedure, based for the most part on our model, that accounts for both
thermal and load time fluctuations. Its application typically entails a preliminary transient thermal analysis, to be followed by a system identification procedure provided
the necessary field measurements are available. The feasibility of this methodology
has been illustrated through an analysis of three different structures. To the best of

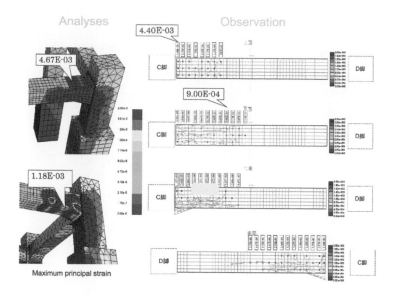

FIGURE 7.43   Close-up of analysis 4: Strains

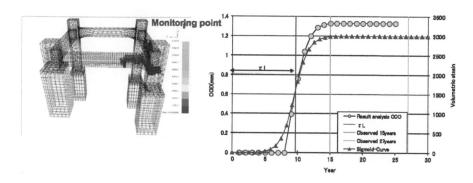

FIGURE 7.44   Kinetics of analysis 4

the author's knowledge, this approach represents the most comprehensive analytical methodology for investigating AAR-affected structures to date (based upon a realistic assessment of the investigated structures) and will hopefully stimulate further research and applications along these lines.

8 Micro Model

This chapter constitutes a departure from the main theme of the book as if will focus exclusively on micro-models. Though certainly not essential to the mere undertaking of a structural analysis, it does provide the interested reader with a greater insight on the underpinning chemical reaction being modeled.

The chapter is composed of two distinct parts. The first is an attempt to simulate the expansion of a reactive mortar bar through the diffusion of ions and gel. Diffusion is simulated by finite difference, and expansion through finite element. In the process, an attempt is made to correlate the resulting time dependent expansion with the sigmoidal curve of Larive. The second part is again a timid attempt to develop a mathematical model for the kinetics of the AAR. To the best of the author's knowledge, this has never been attempted before.

In both cases, attempts will be made to predict the kinetics curve of Larive from either a diffusion formulation, or by examining the kinetics of the chemical alkali silica reactions.

8.1 A DIFFUSION-BASED MICRO MODEL

Written in collaboration with Prof. Wiwat Puatatsananon

To the best of the author's knowledge, the first report on a coupled chemo-mechanical simulation of AAR was released by Sellier et al. (1995). While not explicitly modeling both phases of the AAR reaction, Ulm et al. (2000) presented a model for AAR swelling formulated at the micro level, which was subsequently applied at the structural level. A micro-mechanical model based on extended finite element modeling (XFEM) was developed by Dunant and Scrivener (2010); however, this model did not attempt to explicitly simulate diffusion, but only the resulting swelling.

Microscopic chemo-mechanical models are able to investigate the heart of the reaction since they simulate both diffusion and expansion. The first such model was developed by Suwito et al. (1998) and later expanded by (Suwito et al., 2002). In this model, the diffusion of chemical ions from the pore solution into aggregate and the permeation of AAR gel from the aggregate surface into the surrounding porous cement matrix are both simulated. The total AAR gel is divided into two parts: the gel directly deposited in interface pores, which does not cause expansion; and the gel permeating into the surrounding pores in cement paste, which generates interface pressure and is responsible for expansion. Interface pressure between aggregate and cement paste has also been accounted for herein. The model is placed within the framework of the self-consistent method, and an analytical solution (albeit for spherical aggregates) has been derived. This model ultimately provides an effective expansion coefficient in terms of aggregate fineness.

A similar model was subsequently developed by Poyet et al. (2007), where interface pressures are not explicitly included, though isotropic damage caused by AAR expansion and its resulting swelling have both been quantified. This model was capable of capturing the reaction kinetics and suitable for application to a system of several aggregates, prior to being refined by Multon et al. (2009).

Jin et al. (2000) investigated the feasibility of "Glascrete", whereby (silica-rich) waste glass is used as aggregate. An assessment was made through a series of C-1260 mortar bar tests, according to which 10% of innocuous sand was replaced by glass particles of varying sizes. These authors showed that a decrease in particle size causes greater volume expansion and damage due to AAR, up to a given particle size (worst case). However, as particle size was reduced further, smaller AAR-induced expansions were observed, along with an increase in concrete compressive strength.

The main objective of this chapter is to present the numerical simulation of the ASTM C-1260 test, using a method inspired by Suwito et al. (2002).

ASTM C-1260 (ASTM C1260, 2000) is an accelerated test to detect the potential of an aggregate intended for use in a concrete specimen undergoing the alkali-silica reaction resulting in potentially deleterious internal expansion. In this test, a 2.54 cm x 2.54 cm x 25.4 cm mortar bar is moisture cured for 1 day prior to demolding and then immersed in a sodium hydroxide solution for 14 days. Lengths are measured daily, and the expansion is assumed to provide an indicator of the aggregate's reactivity. It is stipulated that an expansion of less than 0.1% should be considered innocuous.

This test served as the impetus for Suwito et al. (2002) analytical model, which will be detailed below and placed within a numerical framework. The proposed model will thus remove some of the original limiting assumptions, resulting in a highly nonlinear coupled formulation. The solution will be performed numerically by combining finite differences and finite elements; the former will be used for the diffusion part while the latter will determine the interface pressure.

8.1.1 ANALYTICAL MODEL

In considering a (ASTM C-1260) mortar bar made of non-reactive cement, we are seeking to numerically simulate its expansion by focusing on two opposing diffusion processes that take place simultaneously, namely:

1. Diffusion of the hydroxide and alkali ions into aggregates, followed by the reaction of these ions with the reactive aggregate.
2. Diffusion of the expansive AAR gel from the cement paste-aggregate interface out to the porous interfacial transition zone (ITZ). It is assumed that the volumetric expansion of concrete is initiated only when the volume of reaction product exceeds the ITZ porous volume.

8.1.1.1 Diffusion Models

Three distinct types of diffusion are to be considered: 1) a macro-diffusion of ions from the sodium hydroxide solution, where the bar is immersed, into the mortar bar itself; 2) a micro-diffusion of these ions into the aggregates then occurs; and 3) this is ultimately followed by another reverse micro-diffusion of gel from the aggregates into the cement paste.

8.1.1.1.1 Macro-Ion Diffusion of Alkali

Since the mortar bar is composed of non-reactive cement, ions from the sodium hydroxide where it is immersed (Fig. 8.1) must first diffuse into the specimen for a

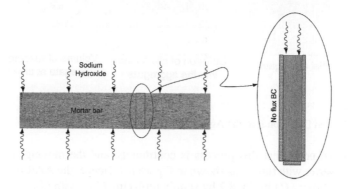

FIGURE 8.1   Macro-diffusion of sodium hydroxide into a non-reactive cement mortar bar

reaction to occur with the silica in the aggregates. It is assumed that the alkali ion causing the gel at the micro level is negligible compared to the alkali ion concentration in the macro level. Hence, the decrease in alkali ion concentration is neglected in the macro-ion diffusion analysis. The macro-diffusion of alkali ions is therefore similar to other macro-diffusion processes in concrete, such as moisture diffusion and chloride penetration, and moreover is governed by Fick's Law :

$$B_{ion,macro}\frac{\partial C_{ion}}{\partial t} = \nabla(D_{ion,macro}(C_{ion})\nabla C_{ion}) \qquad (8.1)$$

where C_{ion} is the free ion concentration of the pore solution inside the concrete, and $B_{ion,macro}$ and $D_{ion,macro}$ are the binding capacity and ion permeability of the concrete, respectively.

8.1.1.1.2 Micro-Ion Diffusion Model of Alkali

In micro-level ion diffusion, the penetration of alkali ions from the pore solution into the aggregate is first simulated. To simplify this problem, it is assumed that AAR occurs only when C_{ion} reaches a critical concentration. The alkali ion causing the gel is thus neglected, and the penetration of alkali ions into the aggregate is also governed by Fick's Law:

$$B_{ion}\frac{\partial C_{ion}}{\partial t} = \nabla(D_{ion}(C_{ion})\nabla C_{ion}) \qquad (8.2)$$

where C_{ion} is the free ion concentration of the pore solution inside the aggregate, and B_{ion} and D_{ion} are the binding capacity and ion permeability of the aggregate, respectively. AAR occurs only when C_{ion} reaches a critical concentration C_{cr}.

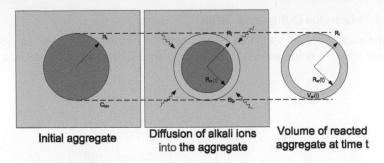

Initial aggregate   Diffusion of alkali ions   Volume of reacted
                           into the aggregate   aggregate at time t

FIGURE 8.2   Ion Diffusion Into the Aggregate

It should be noted that this process is continuous, and the moving front where $C_{ion} = C_{cr}$ varies with time t, as shown in Figure 8.2. Hence, the AAR depth can be solved as a radius $r_{cr}(t)$ in Eq. 8.2 by simply replacing C_{ion} with C_{cr}.

Although Suwito et al. (2002) assumed that D_{ion} is a constant in the linear model, ion diffusivity actually decreases with the formation of gel (mass conservation). A simplified linear model can thus be assumed, i.e.:

$$D_{ion,micro}(t) = \left(1 - b \cdot \frac{C_{gel}(t)}{C_p}\right) D_{ion,micro}^0 \qquad (8.3)$$

where C_p is the porosity of the cement paste, $C_{gel}(t)$ the gel concentration at the aggregate boundary, b a reduction parameter of $D_{ion,micro}$ for the saturated gel state $(0 < b < 1)$, and $b = 0.5$. Due to the dependence of D_{ion} on the ion concentration, Eq. 8.2 becomes a nonlinear equation requiring a numerical solution. It should be observed that the proposed model does not account for the alkali-calcium substitution, as reported by Glasser and Kataoka (1981); this substitution may not appear in accelerated laboratory tests, but instead in the long-term swelling of dams.

For a spherical aggregate, the volume of reactive aggregate of size R_i is given by:

$$V_{R_i}(t) = \left[\frac{R_i^3 - (R_i - r_{cr}(t))^3}{R_i^3}\right]\left(\frac{4}{3}\pi R_i^3\right) = \left[1 - \left(1 - \frac{r_{cr}(t)}{R_i}\right)^3\right]\left(\frac{4}{3}\pi R_i^3\right)$$
$$(8.4)$$

from which the volume of AAR gel can be determined using:

$$V_{gel}^{R_i}(t) = \eta V_{R_i}(t) \qquad (8.5)$$

where η is the volume ratio of AAR gel to the reactive aggregate. $\eta > 1$ means that the volume of reaction product exceeds the original aggregate volume, thus leading to a volume expansion of concrete. η depends on many factors, such as type of aggregate and relative humidity, H. This dependency relationship has been quantified by Poole (1992),

$$\eta = \begin{cases} 0 & H < 50 \\ \frac{H-50}{50}.20\eta_0 & 50 \leqslant H \leqslant 82 \\ \frac{H-82}{18}1.\eta_0 & 82 \leqslant H \leqslant 100 \end{cases} \tag{8.6}$$

where η_0 is the volume ratio when $H = 100\%$. η_0 is set at 1.75 in the present study.

The volume of gel outside the original aggregate boundary will therefore be simply given by:

$$V_{gel}^{*R_i}(t) = V_{gel}^{R_i}(t) - V_{R_i}(t) = (\eta - 1)V_{R_i}(t) \tag{8.7}$$

8.1.1.1.3 Micro-Diffusion of Gel

At the micro level, AAR gel first accumulates along the interfacial zone. Since the porosity of this interfacial zone can only absorb a finite volume of gel before it becomes saturated, any additional gel will generate interfacial pressure. The effective additional gel volume $V_{gel,eff}^{R_i}(t)$ can be determined from:

$$V_{gel,eff}^{R_i}(t) = V_{gel}^{*R_i}(t) - V_{pore}^{R_i} \tag{8.8}$$

where $V_{gel}^{*R_i}$ is the total gel generated outside the original aggregate boundary during the previous process, and $V_{pore}^{R_i}$ is the total volume of pores around the reactive aggregate i (interfacial zone), which is given by:

$$V_{pore}^{R_i} = V_{unit}A_{agg}^{Ri} \tag{8.9}$$

where V_{unit} is a material constant (in the length scale) representing the capacity of the interfacial zone to absorb AAR gel per unit area, and A_{agg}^{Ri} is the surface area of an aggregate particle of size R_i. The gel initially fills the pores (see Fig. 8.3(a)), and once $V_{gel,eff}^{R_i}(t)$ turns positive, the interfacial pressure initiates and drives gel diffusion through the porous cement paste (Fig. 8.3(b)). AAR gel permeation through the porous cement paste can be characterized by Darcy's Law for a viscous flow as :

$$\frac{\partial C_{gel}(t)}{\partial t} = \nabla\left(\frac{K_{gel}}{\eta_{gel}}\nabla P_{gel}(t)\right) \tag{8.10}$$

where $C_{gel}(t)$ and η_{gel} are the gel concentration and viscosity, respectively, K_{gel} is the gel permeability of the porous cement paste, $P_{gel}(t)$ the interfacial pressure distribution due to the AAR gel (which thus depends on the degree of pore saturation). At the aggregate boundary, the interfacial pressure around the aggregate, $P_{int}(t)$ due to AAR gel is applied.

This is very much a coupled chemo-mechanical equation, and its solution relies on a relationship between pressure $P_{gel}(t)$ and gel concentration $C_{gel}(t)$. For the interfacial transition zone (in neglecting gel diffusion into the cement paste), such an equation was originally assumed by Suwito et al. (2002) as:

$$C_{gel}(t) = \beta P_{gel}(t) \tag{8.11}$$

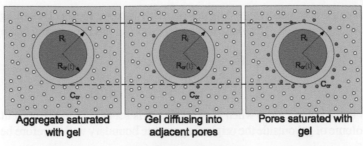

(a) Gel fills the pores first

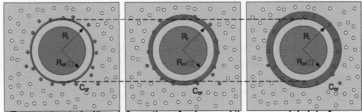

(b) Saturated pores, gel formation in the ITZ

FIGURE 8.3   Micro-diffusion of gel

where β is the state function for the cement paste. When the pores of the cement paste are saturated with AAR gel, $C_{gel} = C_p$, where C_p is the porosity of the cement paste. At the same time, pressure P_{gel} reaches saturation pressure, which can be approximated by the concrete tensile strength f_t'. Therefore: $\beta = C_p/f_t'$. Once the interfacial pressure due to AAR gel around the aggregate $P_{int}(t)$ is determined through a stress analysis, gel concentration at the aggregate boundary can be derived from $C_{gel}(t) = \beta P_{int}(t)$. Hence, Eq. 8.10 can be expressed in terms of C_{gel} and solved numerically. It should be noted that Eq. 8.11 assumes neglecting the gel absorbed by the cement paste, which may result in an artificially increased interfacial pressure between the aggregate and the cement paste (since its concentration has been overestimated). An alternative equation for the pressure induced by the gel is given by Sellier et al. (1995).

The total volume change due to AAR in all aggregates of size R_i is equal to the volume of AAR gel beyond the accommodating capacity of the corresponding interfacial zone. This volume change can be expressed as: $\Delta V_{gel}^{R_i}(t) = V_{gel,eff}^{R_i}(t) - V_{pg}(t)$, where $V_{pg}(t)$ is the total gel permeating into the cement paste, and $\Delta V_{gel}^{R_i}(t)$ will cause the transient internal pressure between aggregate and cement paste matrix, $P_{int}(t)$.

Gel volume in the porous cement paste surrounding the aggregate at time t, $V_{gp}^{R_i}$, is given by:

$$V_{pg}^{R_i}(t) = \begin{cases} \int_{R_i}^{R_\infty} 4\pi r(t)^2 C_{gel}(t) dr & \text{3D spherical particle} \\ \int_{R_i}^{R_\infty} 2\pi r C_{gel}(r,t) & \text{2D cylindrical particle} \end{cases} \quad (8.12)$$

where C_{gel} is the total gel concentration (solution to Equation 8.10). This solution must be obtained either analytically or numerically.

The coefficient of AAR expansion for an aggregate with radius R_i can then be determined from:

$$\alpha_i^{R_i}(t) = \frac{\Delta V_{gel}^{R_i}(t)}{V_a^{R_i}} = \frac{V_{gel,eff}^{R_i}(t) - V_{pg}^{R_i}(t)}{V_a^{R_i}} = \frac{V_{gel}^{R_i}(t) - V_{pore}^{R_i} - V_{pg}^{R_i}(t)}{V_a^{R_i}}$$

$$(8.13)$$

This equation assumes that aggregate swelling is solely caused by the formation of gel outside the interfacial zone (as gel inside the pores does not cause swelling). Given $\Delta V_{gel}^{R_i}(t)$, $P_{int}(t)$ can be solved from the equilibrium of the composite system, which becomes a mechanics problem for interfacial stress analysis in either a three-dimensional (3D) spherical system or a two-dimensional (2D) cylindrical system.

Finally, the total coefficient of expansion of an AAR-affected concrete equals the sum of all individual aggregate expansions (each one dependent on its size). The interfacial pressure determined in this manner must be equal to the pressure obtained from the state equation used in 8.10, which is clearly a nonlinear formulation that can only be solved iteratively.

8.1.2 NUMERICAL MODEL

The previously described analytical model, which is an extension of the model by Suwito et al. (2002), can only be solved numerically since it entails the solution to coupled nonlinear equations, which will be derived through three distinct steps: 1) a macro-ion diffusion analysis, 2) a micro-coupled, chemo-mechanical gel diffusion analysis, and 3) a macro stress analysis. This procedure enables determining the level of mortar bar expansion (Fig. 8.4). Furthermore, since a numerical solution is being

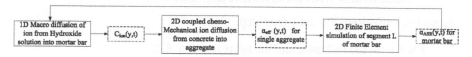

FIGURE 8.4   Numerical procedure for AAR expansion determination

derived, the "circular" aggregate constraint is no longer pertinent.

8.1.2.1 Macro-Ion Diffusion Analysis

The diffusion analysis of ions from the concrete boundary into the concrete is performed through a finite difference scheme by using the heat transfer analysis code from the concrete deterioration analysis program (CDAP) (Puatatsananon, 2002). This macro-ion diffusion analysis is not coupled with the micro-ion diffusion analysis due to the lack of a term for the alkali ion concentration consumed by the gel creation reaction in Eq. 8.1. The interfacial pressure between aggregate and cement paste increases whenever the gel formation rate exceeds the gel diffusion rate. On the other hand, the pressure drops whenever the gel diffusion rate is greater than the gel formation rate. For small aggregate sizes therefore, gel formation can result in high interfacial pressure (high gel concentration at the aggregate boundary) during the first time step and then cause an increased gel diffusion rate and interfacial pressure drops during the next time step. The temporal variation of the free ion concentration is recorded for the subsequent micro-ion diffusion analysis.

8.1.2.2 Micro-Coupled Chemo-Mechanical Analysis

This analysis is driven by a specially written C++ code (SIMulation of Silica Aggregate Reactions, or SIMSAR). Equations 8.2 and 8.10 (for ion and gel, respectively) are solved by means of a finite difference scheme. The code is coupled with a nonlinear finite element code, (Saouma et al., 2010) for the stress analysis in order to determine the interfacial pressure existing between the aggregate and the cement paste matrix $P_{int}(t)$; this analysis is performed only for the RVE of concrete, which contains the reactive aggregate inside. The gel concentration at the reactive aggregate boundary is then determined from $C_{gel}(t) = \beta P_{int}(t)$, and Eq. 8.11 can be solved numerically using a finite difference scheme.

The numerical simulation at a given time step t of the ion diffusion from the sodium hydroxide solution can be described as follows:

Initialization prior to the analysis:

1. A topological representation of both the aggregates (as identified by the subscript i) and cement paste is defined first, followed by setting up a corresponding finite difference grid.
2. Initial data consist of the: critical ion concentration C_{cr}, porosity C_p, ion permeability of the aggregate $D^0_{ion,micro}$, ion binding capacity B_{ion}, and volume ratio of AAR gel to reactive aggregate η.

Micro-Ion Diffusion: For each time step

1. Solve for ion diffusion into the aggregate; Eq. 8.2.
2. For each node, update $D_{ion}(t)$; Eq. 8.3.
3. For each aggregate i
   a. Count the number of cells with $C_{ion} \geqslant C_{cr}$, and determine the volume of reactive aggregate $V_{R_i}(t)$.
   b. Determine $V^{R_i}_{gel}(t)$ from Eq. 8.5, and $V^{R_i}_{gel,eff}(t)$ from Eq. 8.8.
   c. If $V^{R_i}_{gel,eff}(t) > 0$, set $t_{gstart} = t$ and then proceed to the next step; otherwise, increase time step Δt and repeat the ion diffusion simulation.

Gel Diffusion with $t_{total} = t_{gstart} - t_{istart}$, $t_{gel} = t_{istart}$ and $\Delta t_{gel} = \Delta t$. Assuming $V_{pg} = 0$ for $t_{gel} = t_{gel} + \Delta t_{gel}$, and for each time increment

1. Determine total expansion coefficient for each aggregate α_T^{Ri} from time $t = t_{istart}$ to $t = t_{gstart}$ using Eq. 8.13.
2. Determine the expansion coefficient for each gel diffusion time increment Δt_{gel} $\alpha_{inc}^{Ri} = (t_{gel} - t_{istart})\alpha_T^{Ri}/t_{total}$.
3. Input α_{inc}^{Ri} for all aggregates into MERLIN for the stress analysis.
4. MERLIN will then perform the interfacial stress analysis by means of a pseudo-thermal analysis. Given the propensity for microcracking around the aggregates due to localized high pressure, the stress analysis is nonlinear with a fracture-plasticity constitutive model (Červenka and Červenka, 1999).

   a. Apply α_{inc}^{Ri} as $\alpha_{inc}^{Ri} = \alpha_i^T \Delta T$ where ΔT is a virtual unit temperature change for the aggregates, and α_i^T will be the pseudo-coefficient of the thermal expansion for each particle.
   b. Analyze the problem
   c. Return the traction distribution around each aggregate.
   d. Return the vertical and lateral displacements.
5. From the traction distribution around each aggregate, determine the interfacial pressure. Gel concentration at the aggregate boundary is then derived from $C_{gel}(t) = \beta P_{int}(t)$. Note that this method only yields the interfacial pressure between aggregate and cement paste.
6. For gel diffusion into the cement paste, solve Eq. 8.10 and 8.11.
7. Count the number of nodes with a nonzero gel concentration and determine the volume of gel in the porous cement at time t_n in the vicinity of particle i, $V_{pg}^{Ri}(t)$ (which essentially entails solving Eq. 8.12).
8. Solve for $\alpha_i^{Ri} = \alpha_{inc}^{Ri} - \dfrac{V_{pg}}{V_a}$
9. Compare α_i^{Ri} with α_{inc}^{Ri}. If their difference lies within the prescribed tolerance, increase the gel diffusion time step and perform the analysis during the next gel diffusion increment; otherwise, store the value of $\alpha_{inc}^{Ri} = \alpha_i^{Ri}$ and then input α_{inc}^{Ri} back into MERLIN until convergence is achieved.
10. If $t_{gel} = t_{gstart}$, determine the isotropic or anisotropic (if the RVE has a different confining traction in the vertical and horizontal directions) AAR-induced strain.

$$\varepsilon_x^{AAR}(t) = \frac{u_x}{\text{Width}} \tag{8.14}$$

$$\varepsilon_y^{AAR}(t) = \frac{u_y}{\text{Height}} \tag{8.15}$$

Set the larger value of the two strains as the expansion coefficient due to AAR gel in the particular RVE ($\alpha_{RVE}(t)$); next, increase time step t and return to the ion diffusion analysis. The anisotropic loading is not assumed to alter the gel pressure.

The algorithm for the coupled chemo-mechanical simulation of gel diffusion into the aggregate is shown in Fig. 8.5. In sum, the major differences between the original analytical model by Suwito et al. (2002) and the proposed numerical model are as follows:

1. The numerical model can handle multiple and non-spherical aggregates.
2. The interfacial pressure distribution is more accurately determined through a nonlinear finite element analysis.
3. The effects of humidity and ion concentration are incorporated through Equations 8.6 and 8.3.

8.1.2.3 Macro-Stress Analysis

Once an aggregate expansion has been obtained in terms of elevation y and time t, the overall bar expansion can be sought. A macro finite element mesh, in which one in ten aggregate locations is assumed to be reactive, is to be prepared. At each time step, this expansion is given from the previous analysis while performing a stress analysis with MERLIN. For purposes of simplification, the problem is formulated with just a one-way coupling, and the stress state evaluated in the macro stress analysis is not taken into account at the REV level. Furthermore, the gel cannot permeate outside the REV boundary in the micro gel diffusion analysis, a constraint that in turn will provide the overall mortar bar expansion in terms of time.

8.1.3 EXAMPLE

8.1.3.1 Model

The initial macro ion diffusion from the external hydroxide solution into the concrete (Eq. 8.1) was performed by a pseudo-1D finite difference analysis with 31 nodes across

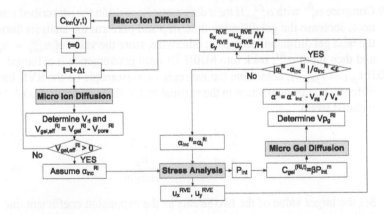

FIGURE 8.5 Algorithm of Simsar and Merlin's coupled simulation

and 31 vertical nodes. The subsequent RVE for the coupled chemo-mechanical model required a total of three meshes. The finite difference mesh for the micro ion diffusion of a single aggregate (i.e. in ignoring the interaction among adjacent aggregates during this step) was set at 620 by 620. The finite element mesh for a single aggregate stress analysis is shown in Figure 8.8(b); this same generic mesh was used for all aggregate sizes by simply scaling the dimensions.

The finite element mesh for a macro analysis intended to yield a global response is shown in Figure 8.6. Given the variation in aggregate size, only a representative segment L of the mortar has been modeled for purposes of the stress analysis (Fig. 8.6(a)). Table 8.1 lists L and the number of elements in the vertical direction for each mesh (all meshes were 12.7 mm high, which corresponds to half the mortar bar height). The RVE size used in the micro-model in Table 8.1 has been determined

TABLE 8.1

Size of RVE used in both the micro and macro level stress analyses; h=12.7 mm

Aggregate size	Diameter [mm]	Size of square RVE in the micro model [mm]	Distance between aggregate and RVE boundary	Selected mortar bar length L [mm]	Number of vertical elements in the FE mesh, n_v
No. 8	2.36	3.335	0.487	126.7	4
No. 16	1.18	1.668	0.244	126.8	8
No. 30	0.60	0.849	0.125	127.4	15
No. 50	0.30	0.426	0.063	126.9	32
No. 100	0.15	0.214	0.032	31.7	64
No. 200	0.075	0.108	0.0167	8.4	128
No. 400	0.038	0.058	0.009	4.1	256

from the aggregate or cement paste volume fraction given in Table 8.2. Figure 8.6(a) shows segment L of the mortar bar being simulated; moreover, Figure 8.6(b) depicts the finite element mesh used for the stress analysis and the RVE for micro-diffusion in one aggregate. For the sake of simplicity, boundary condition compatibility has not been taken into consideration in the macro-stress analysis.

8.1.3.2 Analysis Procedure

The analysis was again conducted according to three major steps:

I Macro-Ion Diffusion of alkali from the sodium hydroxide solution: the concrete is simulated using the finite difference code CDAP program, (Puatatsananon, 2002). $D_{ion,macro}/B_{ion,macro}$ in Eq. 8.1 is set at $1 \times 10^{-11}\ m^2/s$. For the sake of simplicity,

TABLE 8.2

Material parameters used in the diffusion process

Parameter	Value
Maximum time (in days)	14
Step size (in $seconds$)	10,000
Volume fraction of aggregate	0.65
Volume fraction of matrix	0.35
D_{ion}/B_{ion} (mm^2/sec)	1×10^{-10}
ν_{gel} (mm^2/sec)	1×10^{-10}
Critical ion concentration, C_{cr}	0.005
Constant ion concentration, C_o	0.1
Volume ratio η	1.75
V_{unit} (mm^3/mm^2)	0.002
Porosity, $C_p(\%)$	40

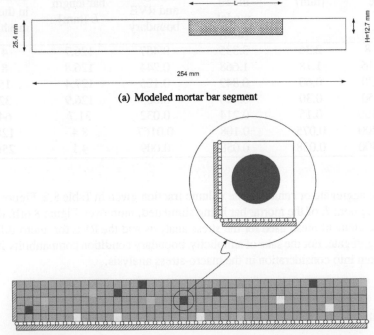

(a) Modeled mortar bar segment

(b) Corresponding finite element mesh for stress analysis, and RVE for the micro-diffusion

FIGURE 8.6   Mortar bar simulation

only a vertical diffusion has been assumed; such an assumption is reasonable given the specimen's large aspect ratio. Ion concentration of the solution equals 0.1, (Suwito et al., 2002), and the spatial variation in the free ion concentration at 14 days is shown in Figure 8.7. Lastly, the spatial and temporal distributions of the free ion concentration, $C_{ion}(y, t)$, has been recorded for subsequent application as a boundary condition for the micro-ion diffusion into the individual aggregate. Note that variations along the mortar bar length (x-direction) have been neglected.

FIGURE 8.7   Free ion concentration of the mortar bar at 14 days

II Micro-Coupled Chemo-Mechanical Analysis will be performed next for each of the aggregate sizes listed in Table 8.1. To enhance computational efficiency, the coupled micro simulation is only performed at four equally spaced elevations in the finite element mesh (Fig. 8.8(a)), as opposed to the n_v elevations indicated in Table 8.1. Intermediate results at other elevations have simply been interpolated, in a step driven by the high computational time required for this coupled analysis. Let's note that such an analysis will yield the coefficient of expansion at predefined elevations and given times. The procedure adopted is as follows:

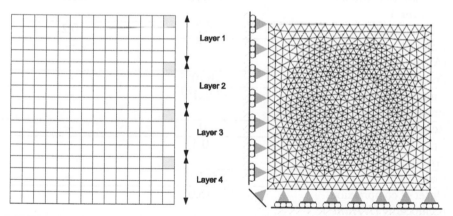

(a) Finite difference mesh for the coupled macro-   (b) Finite element mesh for a single aggregate
analysis   RVE

FIGURE 8.8   Numerical discretizations

    1.Subdivide the mortar bar RVE into four layers and determine the ion concentration histogram at the given level from the preliminary macro-diffusion analysis (Fig. 8.8(a)). This free ion concentration will constitute

the ion boundary condition at the aggregate surface in the micro-ion diffusion analysis.

2. Perform the coupled chemo-mechanical analysis of the selected RVE in
   each layer (Fig. 8.8(b)). The selected parameters used in the micro-ion
   diffusion analyses (Suwito et al., 2002) and those introduced in the stress
   analysis are presented in Tables 8.2 and 8.3, respectively. It should be noted
   that in the MERLIN analysis, interfacial elements were used around each
   particle; moreover, a nonlinear continuum element was used for the concrete, thus allowing for the formation of either smeared cracks or crushing
   of the concrete. This procedural step represents another new development
   compared with the analytical solution associated with Suwito et al. (2002).
   No attempt has been made to reconcile the displacement (or traction) compatibility in the RVE analysis (Fig. 8.8(b)) with that in the macro-analysis
   (Fig. 8.6(b)).

TABLE 8.3

Material parameters used in the stress analysis

Parameter	Value	
	Cement Paste	Aggregate
Element group	Plastic fracture model	Isotropic linear elastic
Modulus of elasticity (N/mm^2)	1.2×10^4	8×10^4
Poisson's ratio	0.2	0.2
Uniaxial tensile strength (N/mm^2)	3.0	—
Fracture energy (N/mm)	7.0×10^{-2}	—
Uniaxial compressive strength (N/mm^2)	-3.0×10^1	—
Return direction factor	0	—
Critical compressive displacement (mm)	-0.5	—
Failure surface roundness factor	5.5×10^{-1}	—
Onset of compression nonlinearity (N/mm^2)	-15.0	—
Plastic strain at f'c	-8×10^{-4}	—

3. Compute the expansion coefficient due to AAR gel of the selected RVE
   after 14 days for each of the four layers.
4. Through a regression analysis, determine an equation that characterizes
   the AAR expansion coefficient in terms of distance from the top of the
   mortar bar, Fig. 8.9.

Macro stress analysis: It was assumed that only 10% of the aggregates are reactive. At first, these are randomly located in the finite element mesh and then assigned a pseudo-thermal expansion governed by the previously obtained linear regression (Fig. 8.11). A linear elastic finite element analysis of the entire mortar bar model was then performed (a nonlinear analysis was deemed unnecessary since no major cracks had formed); also, the AAR-induced expansion was determined from the point of maximum displacement (Fig. 8.12).

8.1.3.3   Investigation Results

8.1.3.3.1   Micro-Modeling

For starters, an example of an AAR mortar bar investigation using 10% No. 30 reactive aggregate will be described. In the micro-level analysis, only the top-layer RVE results (for the mortar bar) are shown. Figure 8.10 reveals the development of interfacial pressure around the reactive aggregate over a time period lasting up to 14 days.

At the outset, the gel formation rate is faster than the cement paste diffusion capacity around the aggregate, thus pressure gradually increases around the aggregate. Once this pressure reaches the tensile strength of the cement paste (3.0 MPa), microcracks begin to form, thereby releasing the pressure increase by allowing the gel to diffuse even more. Since the gel formation rate remains higher than the rate of gel diffusion into the cement paste around the aggregate, the pressure may exceed the tensile strength of the cement paste, as illustrated in Figure 8.10.

AAR expansion starts immediately once the gel has filled the porous zone around the aggregate and then permeated into interconnected pores in the cement paste matrix. When AAR expansion around the aggregate has been initiated, both the reaction force at the support and RVE displacement become noticeable, gradually increasing over

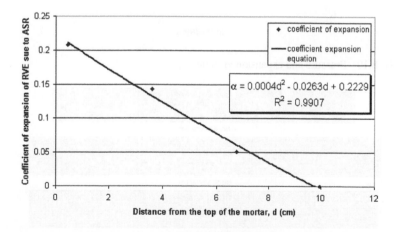

FIGURE 8.9   Linear Regression for AAR Expansion vs. depth

time in the x direction. After $t = 0.2$ days, AAR expansion produces no further change in the reaction at the support, but the RVE displacement still increases over time, a situation that may be due to the strain-softening behavior of concrete. Once strain softening has occurred, the reaction at the support in the x direction remains constant while the RVE strain keeps increasing.

After investigating the coefficient of expansion due to AAR gel on the No. 30 (0.6-mm) reactive RVE of the top element in all four layers considered (as shown in Fig. 8.8(a)) at 14 days, the expansion coefficients of these four points are plotted on the graph along the depth from the top of the bar, and the expansion coefficient equation is derived in accordance with Figure 8.9.

At the macro level, the The random RVE locations of the 10% reactive aggregate inside the mortar bar are shown in Figure 8.11. The inhomogeneous macro-deformation corresponding to the No. 30 aggregate will induce an anisotropic deformation (see Fig. 8.12).

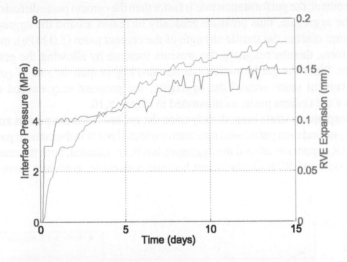

FIGURE 8.10   Pressure and expansion *vs.* time

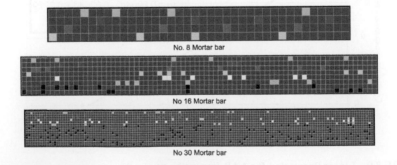

FIGURE 8.11   Random locations of the expansive aggregates

FIGURE 8.12   AAR-induced inhomogeneous macro-deformation

The expansion coefficient of each aggregate size with respect to time is shown in Figure 8.13.

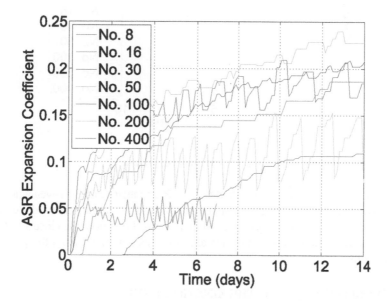

FIGURE 8.13   Simulated coefficients of AAR expansion vs. time relative to aggregate size

This figure reveals an unanticipated oscillation for aggregates 100 or smaller in size. Such an oscillation can be explained by revisiting Eq. 8.13, which yields the AAR expansion coefficient and is linearly dependent on the volume of non-absorbed gel. However, both the ion diffusion and gel generation (for the reactive aggregates) depend on the actual aggregate size. In contrast, gel diffusion depends on the distance between aggregates and the RVE boundary listed in Table 8.1. The interfacial pressure between aggregate and cement paste increases when the gel formation rate exceeds the gel diffusion rate. Moreover, pressure drops whenever the gel diffusion rate is greater than the gel formation rate. For small aggregate sizes, gel formation can therefore lead to a high interfacial pressure (with high gel concentration at the aggregate boundary) during the initial time step; afterwards, gel formation causes an increase in the gel diffusion rate while the interfacial pressure drops during the subsequent time step. Figure 8.14 shows the temporal variation in the oscillation of non-absorbed gel volume beginning for size 100 aggregates and smaller.

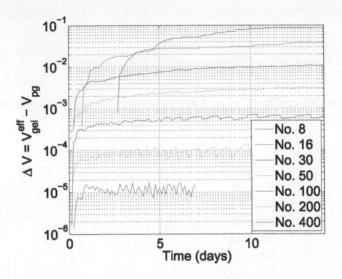

FIGURE 8.14   Temporal variation in non-absorbed gel volume

As a final step, the numerically simulated AAR expansions of the seven aggregate sizes are compared with both experimental results and the numerical model derived from Suwito et al. (2002) in Figure 8.15 after 14 days. Let's note the close similarity between experimentally obtained results and findings generated from the proposed numerical simulation.

8.1.4   FROM DIFFUSION TO THE KINETIC CURVE

8.1.4.1   Preliminary Model

This study will next focus on the expansion curves for aggregate sizes No. 8 (0.84 mm). Eq. 8.37 is based on the maximum expansion, $\varepsilon_{AAR}^{\infty}$, and on two temporal parameters, i.e. latency time (τ_L) and characteristic time (τ_C) (Fig. 8.16). By properly adjusting these three parameters, practically any expansion can be reproduced, including cases where the reaction is continuous. This set-up recognizes the thermodynamic nature of the reaction through an Arrhenius type equation (Arrhenius, 1889), where θ is the actual absolute temperature, and θ_0 the absolute temperature corresponding to τ_L and τ_C. Because these experimental tests, as reported by (Jin et al., 2000), were conducted at 80°C for 14 days, Equation 8.37 has been superimposed on the model prediction with $\varepsilon_{AAR}^{\infty}$, τ_L and τ_C corresponding to 0.108, 4.5 days and 2 days, respectively. Figure 8.16(b) shows the superposition of Larive's curve-fitted model that matches the expansion kinetics obtained from the numerical model presented. Since an adequate correlation has been obtained using reasonable parameter values, it can be concluded that through proper refinements of our numerical model, AAR expansion kinetics can

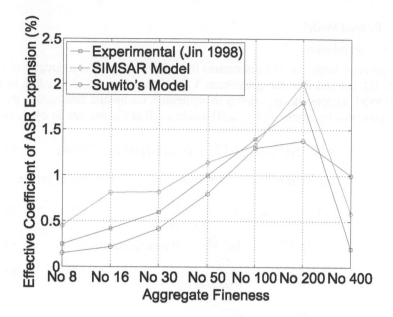

FIGURE 8.15   Expansion vs. fineness

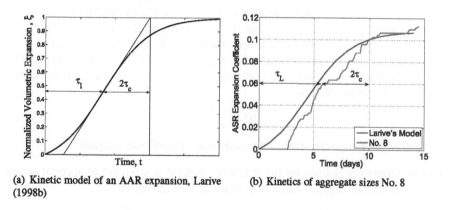

(a) Kinetic model of an AAR expansion, Larive
(1998b)

(b) Kinetics of aggregate sizes No. 8

FIGURE 8.16   Kinetics of the AAR expansion

indeed be determined and input into the macro-analysis of massive structures such as dams, Saouma et al. (2007).

The procedure described in conjunction with Figure 8.9 can then be applied via a finite element analysis. Figure 8.12 shows the deformed shape, which turns out to be irregular due to the random distribution of the reactive aggregates.

8.1.4.2 Refined Model

8.1.4.2.1 Formulation

During previous steps, we first determined the volume of reactive aggregate of size R_i from Eq. 8.4 and then the gel volume $V_{fel}^{R_i}(t)$ from Eq. 8.5. According to this refined model, an opposite approach is implemented, i.e. by first determining the gel volume generated by aggregate $V_{gel,in}(t)$ inside a cell of dimensions dx by dy from:

$$\frac{\partial V_{gel,in}}{\partial t} = [\alpha_1 . f_1(V_{gel,in}) . f_2(T) . f_3(C_{ion}) . f_4(C_{silica})] dx . dy \qquad (8.16)$$

where α_1 is a non-dimensional parameter relating gel concentration to gel volume.

$$f_1(V_{gel,in}) \quad = \quad (1 - \frac{V_{gel,in}}{V_{gel,max}})^m, \qquad m = 1.0 \qquad (8.17)$$

$$f_2(T) \quad = \quad Ae^{-\frac{E_a}{RT}} \quad \text{Arrhenius equation} \qquad (8.18)$$

$$f_3(C_{ion}) \quad = \quad \frac{C_{ion}}{C_{ion,max}} \qquad (8.19)$$

$$f_4(C_{silica}) \quad = \quad \frac{C_{silica}}{C_{silica,max}} \qquad (8.20)$$

where $V_{gel,max} = dx . dy$ is the gel volume inside a cell, and C_{ion} and $C_{ion,max}$, C_{silica} and $C_{silica,max}$ the actual and maximum alkali and silica concentrations at a specific grid point inside an aggregate. Eq. 8.17 reflects the fact that the generated gel volume cannot exceed the gel volume ($V_{gel,max}$). Eq. 8.18 is simply the Arrhenius Law, where A is a so-called "pre-exponential factor", E_a the activation energy, R the universal gas constant, and T the absolute temperature. Eqs. 8.19 and 8.20 reflect the fact that the higher the reagent concentration, then the higher the gel formation rate. Let's now determine R_i from:

$$V_{R_i}(t) = \sum V_{gel,in} \qquad (8.21)$$

Similarly, we can now reassess the reduction rates of both reagents.

Alkali: Originally defined by Eq. 8.2, this equation is modified to account for the drop in ions consumed in order to produce the gel as follows:

$$B_{ion}\frac{\partial C_{ion}}{\partial t} = \nabla(D_{ion}(C_{ion})\nabla C_{ion}) - \alpha_2\frac{\partial C_{ion,used}}{\partial t} \qquad (8.22)$$

where α_2 is a factor assumed to be equal to unity when silica is present inside the aggregate; otherwise, $\alpha_2 = 0$. $\frac{\partial C_{ion,used}}{\partial t}$ denotes the rate of change of C_{ion} used in the gel reaction inside the aggregate and is given by:

$$\frac{\partial C_{ion,used}}{\partial t} = \alpha_3\frac{\partial V_{gel,in}}{\partial t} \qquad (8.23)$$

where α_3 is a non-dimensional parameter and $\frac{\partial V_{gel,in}}{\partial t}$ the rate of gel volume generated (rate of chemical reaction) at each grid point inside the aggregate.

Silica mass variation inside the aggregate used in the alkali-silica reaction had previously been neglected. Let's now assume it obeys:

$$\frac{\partial C_{silica}}{\partial t} = -\alpha_4 \frac{\partial V_{gel,in}}{\partial t} \tag{8.24}$$

where $\frac{\partial C_{silica}}{\partial t}$ is the rate of silica concentration used in the alkali-silica reaction inside the aggregate. α_4 once again is a non-dimensional parameter.

From a computational standpoint, the proposed procedure is illustrated in Figure 8.17.

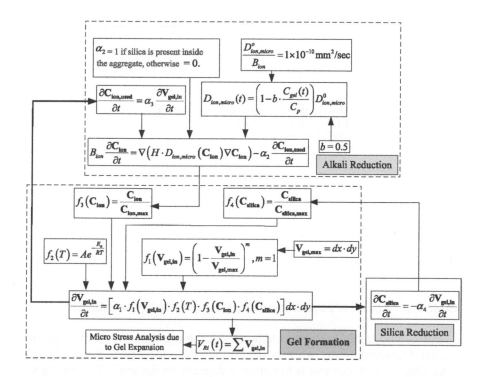

FIGURE 8.17 Flowchart of the refined micro-model for AAR

8.1.4.2.2 Applications

As with all chemical reactions occurring in nature, a limiting reagent is involved, i.e. once the mass of this reagent has been exhausted, the reaction ceases leaving a certain concentration of the other reagent(s). Such a limitation has now been taken into account in this refined model by considering the two possible cases: 1) a finite alkali ion concentration inside the cement paste and an infinite supply of silica concentration inside the aggregate; and 2) an infinite supply of alkali ion concentration inside the cement paste and a finite silica concentration inside the aggregate.

The ensuing analyses, similar to the previous one discussed, are performed by introducing modifications to the model; the results of both these analyses are shown in Figures 8.18 and 8.19, respectively.

Silica>>Alkali: Figure 8.18 reveals that alkali initially diffuses from the cement paste into the aggregate; henceforth, the alkali mass in the cement paste gradually

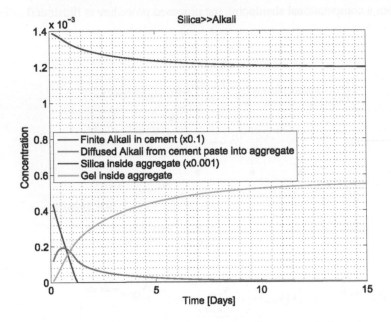

FIGURE 8.18   Kinetics of the AAR reaction for Silica >> Alkali (Excess Silica)

decreases while total alkali concentration inside the aggregate increases along with the gel. More specifically, whereas alkali inside the cement paste is initially considerable, the depletion rate is low since it depends on the level of concentration (Eq. 8.22). Consequently, alkali mass inside the aggregate initially increases until it peaks, at which point its depletion rate (as the gel forms) is higher and thus its total concentration in the cement paste is very low. This low alkali concentration will result in an increasingly lower diffusion into the aggregate, which in turn will cause a rapid decrease in the diffused alkali inside the aggregate. Once the alkali supply has been completely exhausted, the rate of gel generation equals zero.

Alkali>>Silica: Figure 8.19 shows that the rates of alkali diffusion from the cement paste into the aggregate and of gel production are similar. At the same time, the silica is slowly being depleted; when silica concentration becomes very low, the gel concentration freezes.

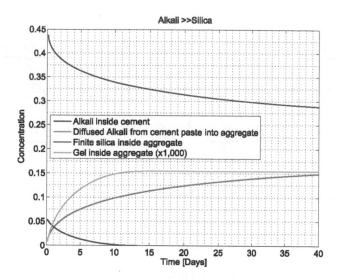

FIGURE 8.19   Kinetics of the AAR reaction for Alkali >> Silica (Excess Alkali)

Let's note that whereas the gel production in Figure 8.18 exhibits the same shape as the curve proposed by Larive, Figure 8.19 more closely resembles the results obtained from mortar bar tests when soaked in alkali solution.

8.2   A MATHEMATICAL MODEL FOR THE KINETICS OF THE ALKALI-SILICA REACTION

Written in collaboration with Ruth A. Martin

At the micro level, it is important to fully understand the kinetics of the chemical reaction and, despite the lack of data of reaction rates, determine what information could be extracted from an insightful reading. In particular, the role of water and the rate of gel formation. Prior familiarization with Appendix C or (Espenson, 2002) may be necessary in order to follow the initial steps of the subsequent derivations. More details can be found in (Saouma et al., 2013). As mentioned in the introduction of this chapter, the ultimate objective herein is to quantify the rates in Equations 1.1, 1.2 and 1.3) and then correlate them with Figure 2.1(a). These equations can be rewritten as follows:

$$\underbrace{Si-O-Si}_{A} + \underbrace{R^+ + OH^-}_{B} \overset{k_1}{\rightarrow} \underbrace{Si-O-R}_{C} + \underbrace{H-O-Si}_{D} \quad (8.25)$$

$$\underbrace{H-O-Si}_{D} + \underbrace{R^+ + OH^-}_{B} \overset{k_2}{\rightarrow} \underbrace{Si-O-R}_{C} + \underbrace{H_2O}_{E} \quad (8.26)$$

$$\underbrace{Si-O-R}_{C} + \underbrace{nH_2O}_{E} \overset{k_3}{\rightarrow} \underbrace{Si-O^- + (H_2O)_n + Na^+}_{F} \quad (8.27)$$

where k_i is the reaction rate. Thus, the rate of concentration change for each of the six constituents can be written (for convenience we will omit the brackets customarily used to refer to the concentration [X]) (Espenson, 2002) as

$$\frac{dA}{dt} = -k_1 AB, \quad (8.28)$$

$$\frac{dB}{dt} = -k_1 AB - k_2 DB, \quad (8.29)$$

$$\frac{dC}{dt} = k_1 AB + k_2 DB - k_3 CE, \quad (8.30)$$

$$\frac{dD}{dt} = k_1 AB - k_2 DB, \quad (8.31)$$

$$\frac{dE}{dt} = k_2 DB - k_3 CE, \quad (8.32)$$

$$\frac{dF}{dt} = k_3 CE, \quad (8.33)$$

where where k_m is a positive real constant for $m = 1, 2, 3$ and X is the concentration of substance X in moles per volume of solution. We seek a solution to this system of coupled six ordinary differential equations, and the temporal evolution of F (expanded gel) in particular.

It can be shown (Saouma et al., 2013) that the solution of this system of equations reduces to solving

$$A'' = \frac{A'}{A}\left[\left(1 + \frac{k_2}{k_1}\right) A' + (2k_2 - k_1) A^2 - M_1 k_2 A\right] \quad (8.34)$$

$$B'' = \frac{1}{B}\left[(B')^2 - (k_1 + k_2) B' B^2 - k_1 k_2 B^3 (B - M_1)\right] \quad (8.35)$$

$$F' = k_3 \left(M_3 + \frac{A'}{k_1 A} - F\right)\left(M_3 - M_2 + A + \frac{A'}{k_1 A} - F\right)$$

$$= \frac{k_3}{k_1^2 A^2}\left[(M_3 - F) k_1 A + A'\right]\left[(A - M_2)k_1 A + (M_3 - F)k_1 A + A'\right]. \quad (8.36)$$

Thus, one has to solve the set of three equations (8.34), (8.35), or (8.36). Unfortunately, attempts to solve it analytically failed, and a numerical solution is sought instead.

The numerical solution of the original first order equations (8.28)-(8.33) is sought next. More specifically, the behavior of A, B, and D will be examined first, since the evolution equations for these functions can be decoupled from the equations for C, E, and F. Those last concentrations will be determined from the first three and the derived conservation laws. Numerical solutions, based on the Runge-Kutta Method for systems of linear ordinary differential equations (as implemented in MATLAB), will be presented next.

Examining the temporal behavior of F (gel), it can be shown that not only is F always increasing, but $F' \sim 0$ as $t \to \infty$, so that F approaches a positive constant. $F(t)$ is depicted in Figure 8.20(a) In particular, notice that F is larger when more water is present in the system and when the initial concentration of the alkali is larger. That is, F takes on larger values when E_0 and B_0 are increased. Additionally, Figure 8.20(b) depicts the same plot for small times, so that one sees that $F(t)$ has a slower growth rate for very small times. This initial growth rate is even slower for lower values of E_0 since then the growth term in (8.33) is small. Additionally, one should note that the growth rate can be further slowed for small times by decreasing k_1.

The asymptotic behavior of F can be seen in Figures 8.21. Note that the numerical results depicted become more accurate, and agree better with the results if one integrates the equations for longer times. In Figure 8.21, it was assumed that $t = 10$ represented a sufficiently long time to observe the asymptotic behavior of the functions. One could increase this value of t to improve results.

Finally, Figs. 8.20(a) and 8.20(b) should be compared with the macroscopically observed concrete expansion due to ASR by Larive (1998b) as overned by

$$\xi(t, \theta) = \frac{1 - e^{-\frac{t}{\tau_c(\theta)}}}{1 + e^{-\frac{(t - \tau_l(\theta))}{\tau_c(\theta)}}} \tag{8.37}$$

where $\xi(t, \theta)$ is the normalized expansion. and shown in Fig. 8.22. The similarity is worth noting. Future work should go beyond mere qualitative comparison between those two models and attempt to seek a quantitative one.

The preceding model is only a first attempt to improve our fundamental understanding of the ASR. It is hampered by a lack of data (kinetic rates), inconclusive understanding (governing chemical reaction), and some simplifying assumptions (ignring effect of temperature and stoichiometry), yet some interesting results are obtained. More precisely;

1. Confirmed the importance of the initial concentration of water in gel formation.
2. Qualitative similarity (sigmoidal curve) between the concentration of gel (chemical reaction) and the expansion of concrete under ASR (physical macroscopic laboratory observations)

More details and results are reported in Saouma et al. (2013).

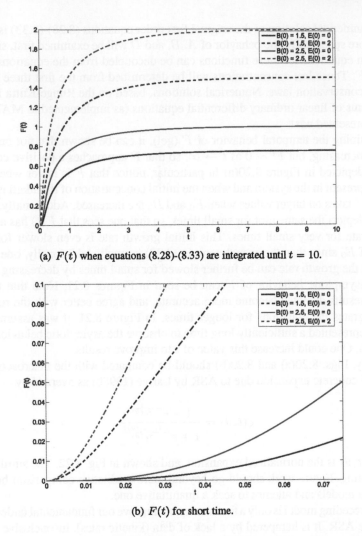

(a) $F(t)$ when equations (8.28)-(8.33) are integrated until $t = 10$.

(b) $F(t)$ for short time.

FIGURE 8.20 F as a function of time for various initial conditions. The parameters are $k_1 = 1, k_2 = 10^6$, and $k_3 = 50$. The other initial conditions are $A(0) = 1, C(0) = D(0) = F(0) = 0$.

8.3 SUMMARY

The chapter has first expanded the mathematical model of Xi by casting the formulation in a coupled finite difference - finite element framework which did also allow for a generalisation of the model. Expansion in terms of aggregate size is addressed, and the actual expansion of a mortar bar simulated. Interestingly, it was determined that, qualitatively, the proposed model could also yield a sigmoidal expansion analogous to the one experimentally observed by larive.

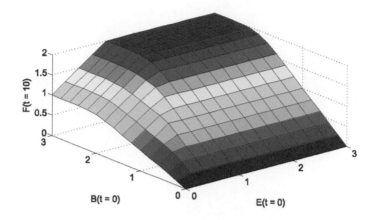

FIGURE 8.21 $F(t = 10)$ as a function of $B(t = 0)$ and $E(t = 0)$.

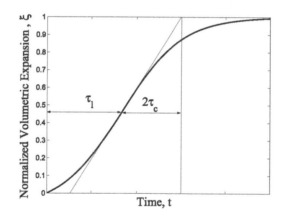

FIGURE 8.22   Theoretically predicted ASR expansion (Larive, 1998b)

The second part was an attempt to use tools used by Chemists derive a mathematical model for the kinetics of the AAR. In the absence of laboratory data, assumptions were made as to the rate of individual expansions, and qualitative observation made. It is hoped that this innovative work will be of interest to young researchers who could further refine it.

9 Prediction of Residual Expansion

Chapter written in collaboration with Mohammad Amin Hariri-Ardebili

Once AAR is known to occur, two questions take on paramount importance: 1) How will AAR affect structural integrity and safety? and 2) For how long will the reaction continue? The answer to the first is more straightforward and undoubtedly entails a "proper finite element simulation", as discussed in the previous chapters. The second question, on the other hand, proves to be very difficult to address and, not surprisingly, has received (comparatively) very little attention from the research community. This chapter will nevertheless seek a response to this second question by first reviewing the (rather limited) body of literature in detail and then presenting a methodology, though not yet fully developed, capable of providing a partial answer.

9.1 LITERATURE SURVEY

Three models will be reviewed herein, though only the third explicitly addresses determining the potential for future expansion.

9.1.1 ESTIMATION OF PREVIOUS AAR EXPANSION, BERUBE ET AL. (2005)

To quantitatively assess previous AAR-based expansion in an actual concrete structure, Smaoui et al. (2004) indicated three potential tests:

Surface cracking (or Accumulated Surface Cracking, ASC)[1] resulting from AAR (once both shrinkage cracks and freeze and thaw cracks have been removed from consideration) using one of two methods:

    1.Institution of Structural Engineers (1992), which consists of measuring the width of cracks intersecting at least five parallel lines 1 m long, traced along the concrete and separated by at least 250 mm from one another. The total deformation due to AAR (and other deleterious mechanisms) is assumed to be equal to the sum of the widths of all intersected cracks divided by the total length of the reference lines.

    2.*Laboratoire Central des Ponts et Chaussées* (LCPC, 1997), which consists of measuring the width of all cracks intersecting two perpendicular lines

---

[1]This has been referred to as Crack Index in Sect. 1.2.5.

1 m long originating from the same point and their 1.4-m long diagonals drawn at the surface of the concrete element under study.

These measurements are nondestructive, inexpensive, easy to perform and repeatable over time at the same location. They may be carried out with either a simple crack width ruler (Fig. 9.1(a)) or an electronic ruler (Fig. 9.1(b)).

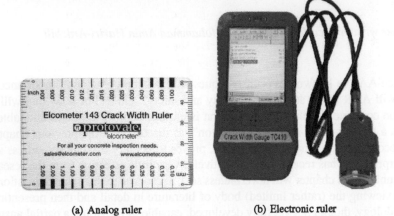

(a) Analog ruler   (b) Electronic ruler

FIGURE 9.1   Field measurements of crack widths

Stiffness Damage Tests (SDT) consists of the cyclic compressive loading of concrete core samples (five cycles) between 0 and 5.5 MPa. It is assumed that both the modulus of elasticity and the hysteresis (area between loading/unloading portions of the stress-strain curves) should be larger during the first loading for cracked concrete than during subsequent loadings. Hence, the energy dissipated during one load-unload cycle (which corresponds to the area of the hysteresis loop) and the accumulated plastic deformation after these cycles are correlated with the closure of preexisting cracks and thus provide a measure of damage in the specimen in the direction of the applied stress.

Damage Rating Index (DRI) is based on a determination by petrographic examination of the internal damage in concrete.

These methods were first assessed with laboratory tests (Smaoui et al., 2004) using non-air-entrained concrete mixes, in accordance with ASTM test method C1293; from each mixture, 20 concrete cylinders, 100 mm by 200 mm in size, were cast. This method was subsequently evaluated (Bérubé et al., 2005) on three existing bridges.

Based on these investigations, it was determined that for existing structures:

1. The measurement of crack widths (ASC) at the surface of the concrete members under investigation appears to be a rather reliable method for estimating the expansion to date as long as the measurements were conducted on the most severely exposed sections of the members; otherwise, the method is capable of significantly underestimating the current expansion.

2. In addition to being more strongly influenced by the petrographer's experience, the DRI did not allow appropriate distinctions to be drawn between the most and least AAR-affected concretes, both visually and mechanically. This study has shown that concretes incorporating the same reactive aggregate may display quite different petrographic features of AAR, depending on whether such features are subjected to natural exposure conditions or accelerated test conditions in the laboratory, which affects the DRI reading.

3. Even if the results are significantly affected by the type of reactive aggregate, SDT appears to be the most effective method for evaluating the expansion to date of the concrete members tested. It seems necessary however to develop a calibration curve corresponding to the type of reactive aggregate under study. Yet this test may underestimate current expansion in the case of a prestressed concrete specimen.

4. In the evaluation of expansion to date in an AAR-affected concrete structure, it is recommended that surface cracking measurements (i.e. crack widths) be taken first on the most severely exposed sections of the structure. The stiffness damage test (SDT) should then be performed on cores sampled from different locations on various AAR-affected members of the structure which have been chosen to illustrate the range of visual distresses observed. Mechanical tests conducted on core samples should help validate the expansion results obtained from the previous two methods.

5. All three concrete structures investigated in this study incorporated a reactive siliceous limestone, and note that different results could be obtained for structures made from other types of reactive aggregates.

9.1.2 VALUE OF ASYMPTOTIC EXPANSION, MULTON *ET AL.* 2008

Multon et al. (2008) investigated the possibility of estimating residual expansion by extracting cores from large laboratory specimens.

To assess this methodology, two concrete mixes (one non-reactive the other reactive) with nearly identical mechanical properties were prepared. The reactive concrete contained a non-reactive sand and a reactive siliceous limestone coarse aggregate, while the reference concrete was exclusively composed of non-reactive aggregates, see Table 9.1.

TABLE 9.1

Reactive (R) and Non-reactive (NR) concrete mixtures (Multon et al., 2008)

		R mix	NR mix
Cement	Kg/m^3	410	410
water/cement	-	0.52	0.5
0-4 mm non-reactive sand	Kg/m^3	620	730
4-20 mm non-reactive coarse aggregate	Kg/m^3	-	1,015
4-20 mm reactive coarse aggregate	Kg/m^3	1,120	-

The concrete specimens were cast in cylinders (ϕ=160 mm, h=320 mm) and prisms (140×140×280 mm), cured for 28 days under watertight aluminum cover at 20°C and about 40% RH. Cores (ϕ=110 mm, h=220 mm) were drilled from the prisms and cylinders 98 days after casting. Following this initial cure, all specimens were underwent expansion under one of three environmental conditions: a) moist air at 100% RH; b) totally immersed in water; or c) under aluminum sealing at 30% RH, with all exposed to 38°C.

Mass and deformation measurements were carried out on all specimens in accordance with the method used by Larive (1998b). For the drilled cores, an expansion similar to that of the "mother" specimen was observed, except for the mass, which seems to vary less for cores stored in air at 100% RH and in water.

The methodology for processing laboratory test results recommended by Fasseu (1997) was adopted herein. A so-called *expansion index* was determined from the slope of residual expansion versus time between 8 and 52 weeks (Fig. 9.2). A deter-

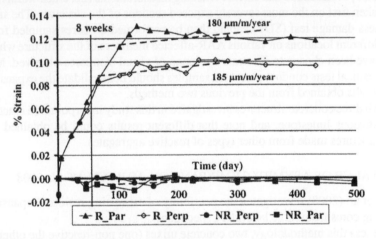

FIGURE 9.2   Rate of expansion between 8 and 52 weeks for cores kept under aluminum at 38°C and 30% RH before coring, (Multon et al., 2004)

mination was made for all six types of cores (two casting directions combined with three initial storage conditions), see Table 9.2. The cores drilled from specimens kept under aluminum before coring show a moderate potential rate of expansion, while all other cores present negligible potential rates of expansion. We have thus concluded that the expansions obtained by the reactive concrete before coring for all three storage conditions are thus fairly low compared to the final expansions recorded in the "mother" specimens; a comparison is provided in Table 9.3. For the cores kept at 100% RH and in water before coring, only half of the final expansion was reached at the time of coring. For those cores sealed with aluminum prior to coring, the measured expansion only represents about 20% of the final value.

The results obtained according to the recommended methodology for data processing by Fasseu (1997) (Fig. 9.2 and Table 9.2) thus appear incapable of predicting

TABLE 9.2

Expansion indices for cores subjected to the residual expansion test (Multon et al., 2008)

previous storage	Direction wrt casting	μm/m/Year	
		Reactive	Non-reactive
Under Aluminum	$\perp$	185	30
	$\parallel$	180	-30
Water	$\perp$	-45	-60
	$\parallel$	90	15
Air at 100% RH	$\perp$	10	40
	$\parallel$	-15	-30

TABLE 9.3

Comparison of expansion at the time of coring and final expansion amplitudes for the reactive concrete (Multon et al., 2008)

	100% RH		Water		Aluminum	
Storage direction	$\parallel$	$\perp$	$\parallel$	$\perp$	$\parallel$	$\perp$
Expansion at coring	0.090	0.058	0.129	0.063	0.022	0.010
Final expansion	0.173	0.120	0.260	0.137	0.116	0.046
% expansion at coring	52%	48%	50%	46%	19%	21%

the actual ultimate expansion measured on the specimens from which the cores were extracted.

In summary, this research has been entirely based on controlled laboratory conditions ($T = 38°C$ and 400 days) and did not present a new methodology for assessing residual expansion in AAR-affected structures, although it did alert us to the difficulty of achieving a correct asymptotic expansion even after one year.

9.1.3 ESTIMATION OF RESIDUAL EXPANSION, SELLIER ET. AL. (2009)

Sellier et al. (2009) developed a methodology for estimating the residual expansion in concrete structures (as opposed to laboratory specimens by Multon et al. (2008)) affected by AAR. A critical (and reasonable) assumption herein is that alkali content is not the limiting factor in AAR, since alkalis may be substituted by calcium. Hence, it is the finite mass of silica that controls termination of the expansion.

9.1.3.1 Preliminary Observations

The initial cores were extracted from a reactive site and then tested for residual expansion at 60°C for one year; it was noted that no asymptotic value could be detected

and moreover the expansion progressed very slowly over time. It was subsequently
determined that the expansion gel produced during accelerated (high temperature)
tests differed from that observed in situ, with this difference being attributed to a
substitution process Na $\leftrightarrow$ Ca, crystallization and a change in molar volume, viscosity
and water content. Accelerated tests therefore may not be suitable for fitting purposes
since they underestimate the swelling rate and do not provide an asymptotic value
(maximum anticipated ε^∞).

Furthermore, the authors note the certain presence of a size effect governing AAR
progression, given its correlation with the surface/volume ratio. Hence, smaller aggregates react faster than larger ones, making it possible to define a kinetics parameter
for each aggregate size (with small aggregates reacting faster than large aggregates).
Aggregates recovered from the site could be crushed first, followed by the preparation
of mortar bars composed of this crushing product. These samples will thus have expe-
rienced extensive preexisting reactivity, resulting in a smaller asymptotic value than
samples obtained from larger aggregates (with little relative expansion). The residual
expansion of a structure can therefore be quantified from such tests.

9.1.3.2   Proposed Procedure

At this point, the authors presented a method for quantifying the kinetic and swelling
amplitude potential at a given site. The essence of this method consists of comparing
residual swelling measured on a mortar made from crushed reactive aggregates with
the residual swelling of mortars made from sound crushed aggregate.

To properly understand this procedure, the model will first be described (note: the
effect of water and RH, though important, will not be addressed in this summary).

It is assumed that a reactive aggregate of volume V_a can create a maximum gel
volume V_g proportional to its own volume:

$$V_g = fV_a \tag{9.1}$$

where f depends on both the reactive silica content of the aggregate and gel texture
(Poyet et al., 2007). The stress-free swelling ε can, in turn, be approximated by:

$$\varepsilon = n^s < V_g - V_p > \tag{9.2}$$

where n^s is the number of reactive aggregates of size s per m^3 of concrete. The final
strain will equal:

$$\varepsilon(\infty) = n < fV_a - V_p > \tag{9.3}$$

where n is the number of reactive aggregates per unit volume of mortar. Since the
reactive aggregates have already undergone partial reaction when extracted from the
structure at time $t = T$, we can then express the residual AAR expansion as:

$$\varepsilon^{res} = n < fV_a(1 - A(t)) - V_p > \tag{9.4}$$

and for $A(t) \simeq 1$, the residual swelling will be close to zero, and conversely for
$A(t) \simeq 0$ the residual swelling will be maximum. The former case corresponds to

"old" concrete and the latter case to "young" concrete. The engineer really needs to know the value of $A(t)$. Another unknown to be determined is f.

The expression for A_T was developed more recently by Gao et al. (2012). It is assumed that the degree of chemical advancement of the reaction (hence gel expansion) is directly related to the degree of silica consumption. For a given aggregate size, we can thus define $A \in [0, 1]$ as:

$$A(t) = \frac{Si^{reacted}(t)}{Si^{reactive}(t = 0)} \tag{9.5}$$

where $Si^{reacted}$ is the mass of silica consumed at time t, and $Si^{reactive}$ the total reactive silica content. As has been well established, chemical advancement depends on the degree of saturation Sr and temperature T; consequently, the derivative of chemical advancement with respect to reaction time t is given by:

$$\frac{\partial A}{\partial t} = \alpha_{ref} \exp\left[-\frac{E_a}{R}\left(\frac{1}{T} - \frac{1}{T_{ref}}\right)\right] \times \frac{Sr - Sr^{threshold}}{1 - Sr^{threshold}}(Sr - A) \tag{9.6}$$

Sr can in turn be determined by numerically solving the nonlinear mass transfer equation. α_{ref} is a constant controlling the kinetics, E_a the AAR activation energy (estimated by the authors at $\neq 47$ kJ/M), R the universal gas constant (8.31 J/M/K), T_{ref} the absolute test temperature, T the current absolute temperature, Sr the degree of saturation of concrete porosity, and $Sr^{threshold}$ the minimum degree of saturation necessary to allow for evolution of the chemical (found to be 2 by Grimal (2007)). Finally, the actual AAR strain can be expressed as:

$$\varepsilon^{AAR} = A(t)\varepsilon^{\infty}) \tag{9.7}$$

and therefore the expansion kinetics ($A(t)$) and final expansion (ε^{∞})) are separately determined.

This method can be broken down into three parts (Fig. 9.3):

1. Field work.
2. Laboratory tests.
3. Inverse finite element simulation.

9.1.3.2.1 Field work

The first component of field work entails typical structural monitoring, which (to the greatest extent possible) should track displacements, cracking, crack opening displacements, temperature and moisture.

The second component consists of core recovery. In the example presented, the structure was composed of two types of concrete, C250 and C350 (with 250 and 350 kg/m^3 of cement respectively), and as is often the case the entire structure was not undergoing AAR expansion. Hence, cores were recovered from each of the two zones, as well as from zones with non-reactive concrete (Steps 1 and 2).

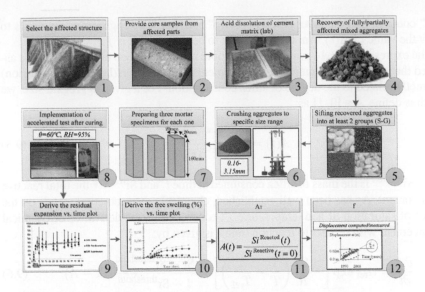

FIGURE 9.3   Schematic description of the Sellier et al. (2009) method

9.1.3.2.2   Laboratory tests

First of all, both intact and degraded aggregates must be very carefully recovered (Steps 3 and 4). A two-step approach is proposed: Coarse aggregates (> 4 mm) are first extracted through heat treatment by immersing the samples in a Na_2SO_4 solution, followed by at least five cycles of freezing and thawing. This procedure will facilitate the extraction of coarse aggregates. Fine aggregates are then extracted through what the authors refer to as an *organic chemical attack* in order to dissolve them from the cement paste. Salicylic acid solution (1.3M/l with methanol as the solvent) is reported to be more efficient than inorganic acid (such as HCl, which may attack reactive silica in the aggregates and thus alter reactivity). Further details regarding chemical extraction and an assessment of chemical advancement can be found in Gao et al. (2013).

The aggregates are then divided into two groups, as shown in Table 9.4 (Step 5). According to this case, $n = 2$ in Eq. 9.3.

The aggregates are then crushed (Step 6) and sifted, with only those in the 0.16 - 3.15 mm range being selected. These selected crushed aggregates are subsequently used to prepare two sets of mixes with 1,500 kg/m^3 sand content, 8 kg/m^3 alkali content and a water-to-cement ratio of 0.5. Next, each mix is cast in at least three mortar specimens (20 × 20 × 160 mm) (Step 7).

The specimens are cured for 28 days in sealed bags at 20°C, before conducting accelerated tests at 60°C and 95% relative humidity (Step 8). Results are plotted as free swelling *vs.* time for each mortar specimen (Steps 9 and 10, Fig. 9.4). These results lead to the following observations:

TABLE 9.4

Mortar bars prepared from crushed aggregate

in-situ Cement Content Kg/m³	Size Range mm	% by weight	Label
250	0-5	32	S250 ◆
	5-16	18	G250 ▪
	16-100	51	
350	0-5	49	S350 ▲
	5-30	51	G350 ×

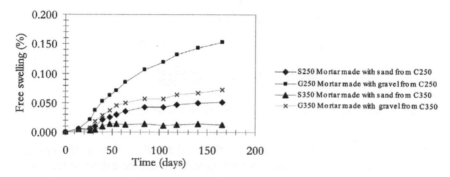

FIGURE 9.4   Swelling of mortar specimens (Sellier et al., 2009)

1. Mortar made from gravel exhibits greater swelling than specimens made from sand. This finding is explained by the fact that *in situ* smaller aggregates (◆ and ×) are more likely than larger ones (▪ and ▲)to have depleted their propensity to react.
2. Swelling capability decreases with cement content of the concrete (generally for a certain percentage of free swelling: *G*250>*G*350 and *S*250>*S*350). Since the amounts of alkali and hydroxyl ions in concrete are proportional to the cement content, the *in situ* reaction of C350 (▲ and ×)concrete was more advanced than for C250 (◆ and ▪); henceforth, in the laboratory, the potential aggregate expansion of C250> *that of* C350.

9.1.3.2.3   Inverse finite element simulation

As mentioned earlier, $A(t)$ (Eq. 9.4) is the primary unknown being sought, along with f (Eq. 9.1). These determinations will be made in the present section.

Silica first reacts *in situ* and then in the laboratory (via the accelerated test, Fig. 9.5).

Two types of aggregate are extracted. Ideally, the intact aggregates are extracted from a dry zone in the structure where AAR has not evolved to a great extent. This

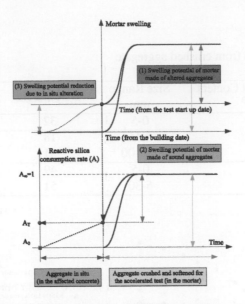

FIGURE 9.5 Swelling of mortar specimens (Sellier et al., 2009)

practice is adopted in order to determine the total reactive silica in the aggregate ($\text{Si}^{\text{reactive}}$). On the other hand, the affected aggregates can be extracted from zones that have clearly been exposed to AAR, allowing the residual reactive silica to be expressed as: $\text{Si}^{\text{residual}} = \text{Si}^{\text{reactive}} - \text{Si}^{\text{reacted}}$.

In assuming that the time of drilling and laboratory testing is t_{lab} relative to construction completion, then the chemical advancement can be expressed as (Step 11):

$$A(t_{\text{lab}}) = \frac{\text{Si}^{\text{reacted}}}{\text{Si}^{\text{reactive}}} = 1 - \frac{\text{Si}^{\text{residual}}}{\text{Si}^{\text{reactive}}} \qquad (9.8)$$

However, since the rate of chemical reaction evolution is independent of the absolute expansion of the structure, it is possible to determine $A(t_{\text{lab}})$ from laboratory experiments and this value would correspond to the amount of silica consumed *in situ*. Hence, integrating Eq. 9.6 yields:

$$\frac{1}{\alpha_{\text{ref}}} = \frac{1}{A(t_{\text{lab}})} \int_{t=0}^{t=t_1} exp\left[-\frac{E_a}{R}\left(\frac{1}{T} - \frac{1}{T_{\text{ref}}}\right)\right] \frac{\text{Sr} - \text{Sr}^{\text{threshold}}}{1 - \text{Sr}^{\text{threshold}}} (\text{Sr} - A)dt \qquad (9.9)$$

With α_{ref} determined, the progress of the chemical reaction can be obtained by revisiting Eq. 9.6.

Lastly, since swelling has been assumed to be directly proportional to the reaction progress, Eq. 9.7, swelling can now be derived from:

$$\frac{\partial \varepsilon^{AAR}(t)}{\partial t} \simeq \frac{\partial A(t)}{\partial t} \varepsilon^{\infty} = \left[\alpha_{\text{ref}} \exp \left[-\frac{E_a}{R} \left(\frac{1}{T} - \frac{1}{T_{\text{ref}}} \right) \right] \right.$$
$$\left. \times \frac{Sr - Sr^{\text{threshold}}}{1 - Sr^{\text{threshold}}} (Sr - A) \right] \varepsilon^{\infty} \quad (9.10)$$

with the only remaining unknown now being ε^{∞}, which is to be determined through curve fitting using *in situ measurements*.

We must still determine f. As mentioned above, f would yield different values for accelerated tests (alkali-silica gel) and slow *in situ* expansion (calcium-silica gel). Henceforth, f will be determined by merely fitting the numerical prediction of structural displacement to the observed displacement (Step 12).

9.2 EXPANSION CURVE FROM DELAYED LABORATORY TESTING

For AAR to occur, high relative humidity must be present, and one of the premises of this book is that AAR has indeed occurred. Furthermore, this chapter is seeking to explore methodologies for ascertaining residual life; "ideally" therefore, a concrete zone would exist with a similar concrete mix yet at low relative humidity. In this case, one could perform accelerated tests (at high temperature in an alkali solution) and extract a quantifiable (albeit approximate) indication of what the final AAR expansion (ε^{∞}) would equal, along with the time required to reach this expansion value.

On the other hand, should a test core not be recoverable, we are then faced with a concrete that has undergone a certain amount of AAR expansion, i.e. enough to result in structural cracks or irreversible deformation, and the critical question then becomes how much additional expansion will occur, and when will this expansion stop. Core testing is indeed possible; however, the test cores would have already undergone a non-negligible AAR expansion and reaction would have already started (for the sake of generalization) at a certain unknown time.

Hence, whether recovered cores are tested in the laboratory or mortar bars prepared according to the procedure outlined in Section 9.1.3.2.2, and with reference to Figure 9.6, we are able to identify three coordinate systems:

t_1-ε_1 associated with actual field conditions, whereby the varying ambient temperature is replaced (as an approximation) by the mean equivalent temperature
t_3-ε_3 associated with laboratory tests at high temperature
t_2-ε_2 associated with the (fictitious) origin of the (backward) extrapolation of the laboratory test at high temperature.

Assuming that the chemical reaction is the same both in the field and at the higher laboratory temperature (this assumption is partially justified by the work of Leemann and Merz (2012)), let's consider concrete expansion in the field and in the laboratory[2].

---

[2]Though different alkali are being mobilized in the two reactions, primarily Na and K in the laboratory, and Ca in the field.

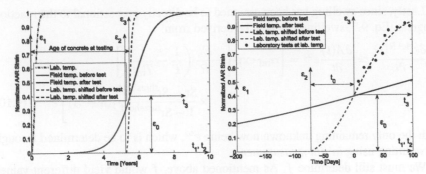

FIGURE 9.6 Nonlinear regression from a delayed laboratory test to determine the full expansion curve (Note the time shift in the second figure)

In the field, expansion starts at $t = 0$ under average temperature $T = T_{field}$. When the expansion has become structurally noticeable (i.e. mapped cracks or irreversible displacements), the core is then extracted (or the mortar bar prepared) and tested in the laboratory at temperature $T = T_{lab}$. At this point, the structure (or, more precisely, the laboratory specimen) has sustained a total expansion of $\varepsilon_0(T_{field})$.

In Figure 9.6, the field expansion (before and after the test) is shown along with ε_0. In the laboratory however, the test is performed at a higher temperature, and the same concrete will now expand much more rapidly. Should we extrapolate the same curve to a zero expansion, the temperature-dependent time t_0 can be defined. Hence, t_0 and ε_0 define the origin of the laboratory coordinate system t_3-ε_3. In turn, t_2-ε_2 defines the coordinate system of the accelerated laboratory test had it been performed soon after the concrete was poured.

In studying this figure closely, let's highlight the fact that the laboratory expansion is indeed accelerated, as the finite data points have been clustered, and the curve is obtained by fitting the data to Eq. 8.37. Finally, it should be noted that t_0 remains unknown.

The following relations are therefore derived:

$$\varepsilon_2(t) = \frac{1 - \exp\left[-\dfrac{t_2}{\tau_c}\right]}{1 + \exp\left(-\dfrac{(t_2 - \boxed{\tau_l})}{\tau_c}\right)} \varepsilon^\infty \qquad (9.11)$$

$$t_2 = \boxed{t_0} + t_3 \qquad (9.12)$$

$$\varepsilon_2 = \boxed{\varepsilon_0} + \varepsilon_3 \qquad (9.13)$$

which would lead to:

$$\varepsilon_3(t_3) = \frac{1 - \exp\left[-\frac{t_3 + t_0}{\boxed{\tau_c}}\right]}{1 + \exp\left(-\frac{(t_3 + \boxed{t_0} - \boxed{\tau_l})}{\boxed{\tau_c}}\right)} \boxed{\varepsilon^\infty} - \boxed{\varepsilon_0} \qquad (9.14)$$

For the sake of clarity, the unknown quantities have been boxed (as in $\boxed{t_0}$). Conceptually, and ideally, we are now in a position to identify the four unknown parameters through a nonlinear regression analysis. Note that in this case, the activation energies U_l and U_c cannot be determined; by now however, they are considered as potential universal constants (Ben Haha, 2006).

9.2.1 NUMERICAL FORMULATION

Following the theoretical derivation of the model, we will now explain how this model can be numerically implemented.

Let's begin by recalling, with reference to Figure 9.7(b), the availability of a set of experimental data points (shown as solid red circles) obtained in the laboratory coordinate system, and the goal now consists of passing the following curve through them.

Hence, given Eq. 9.14 and the laboratory data points, we recast the problem as a nonlinear constrained optimization, with the aim of minimizing the square of errors:

$$\sum_{i=1}^{n} \left(y_i^{\text{Lab}} - y_i^{\text{model}}(\mathbf{x})\right)^2 \qquad (9.15)$$

subjected to the constraints:

$$c_i\left(y_i^{\text{Lab}} - y_i^{\text{model}}(\mathbf{x})\right) \leqslant \varepsilon; 1 \leqslant i \leqslant m \qquad (9.16)$$

$$\left.\frac{d^2 y^{Model}}{dt^2}\right|_{t=\tau_l - t_0} = \varepsilon/10 \qquad (9.17)$$

$$\mathbf{lb} \leqslant \mathbf{x} \leqslant \mathbf{ub} \qquad (9.18)$$

Note that the second constraint implies an inflection point at t corresponding to the latency time, while the last constraint indicates placing a lower bound and an upper bound on the vector $\mathbf{x}$. Thus, Eq. 9.14 can now be rewritten as:

$$y^{\text{model}}(t) = \frac{1 - \exp\left[-\frac{t + x(4)}{x(1)}\right]}{1 + \exp\left[-\frac{(t + x(4) - x(2))}{x(1)}\right]} x(3) - x(5) \qquad (9.19)$$

$\mathbf{x}$ is the unknown vector of the target variables. From Figure 9.6, these variables are:

Model	\multicolumn{5}{c}{At time T_1}
	τ_c	τ_l	ε^∞	ε_0	t_0
Variable	x(1)	x(2)	x(3)	x(4)	x(5)

where n is the number of experimental data points, and $y_i^{\text{model}}(\mathbf{x})$ the analytical equation of the model in terms of the unknown variables $\mathbf{x}$ evaluated at point i.

c is a constraint introduced to force the curve to be close to the laboratory data points, m is the number of constraints, and **lb** and **ub** are the two vectors containing the lower bounds and upper bounds of $\mathbf{x}$.

The second constraint stipulates that the second derivative of the function must be equal to zero at a certain location (corresponding to the inflection point).

9.2.2 ASSESSMENT

In order to test the algorithm, a synthetic dataset is first generated; it will then be assumed that a core is tested at age 5 (based on the laboratory test temperature). Expansion at the two different temperatures is initially shown in Figure 9.7(a), which indicates the difference in time scales for expansion of the same sample at different temperatures. The discrete points (obtained from the synthetic equation) are then fitted through a curve (Fig. 9.7(b)), on which some 15 points from the laboratory tests were used as constraints. Figure 9.7(c) shows the backward extension of the determined curve, while Figure 9.7(d) presents the errors in determining each of the five parameters in the nonlinear regression analysis. The final results of this exercise are displayed in Figure 9.7(e).

The previous results lead to the following observations:

1. It should be kept in mind that all experimental data found in the literature exhibit significant scatter in AAR expansion and moreover they cannot be perfectly fit by Larive's Equation. Hence, a perfect fit in the Matlab regression analysis should not be an expected outcome.
2. The algorithm has now been solidly tested and proven to work successfully.
3. In a nonlinear constrained approximation, it is mathematically impossible to be sure to obtain the absolute minimum (only possible in linear programming, where the objective function and constraints are both linear).
4. Nonlinear optimization is therefore part science and part art; only through the proper variable selection process (acquired through much experience) is it possible to approximate the "exact" solution in a satisfactory manner.
5. In some instances, a small change in input parameter values can considerably improve results.
6. It should be noted that this problem is particularly difficult since only the upper segment of the curve is available, and yet we were still able to recover the entire curve.
7. When the method assesses actual laboratory data, results may not be as smooth.
8. Future work should account for errors (or departure from a highly idealized synthetic curve).

We have thus presented herein the broad outline of a procedure (yet to be fully developed), which could complement other methods, such as the one developed by

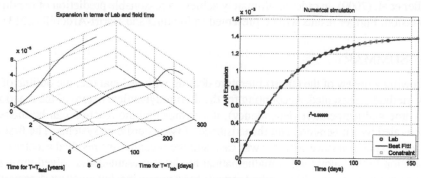

(a) Expansion in terms of both laboratory and field temperatures

(b) Fitted laboratory test results curve (t_3-ε_3 space)

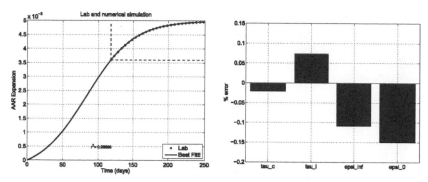

(c) Full curve determined from the laboratory tests (t_2-ε_2 space)

(d) Errors in the estimate of the four unknown parameters

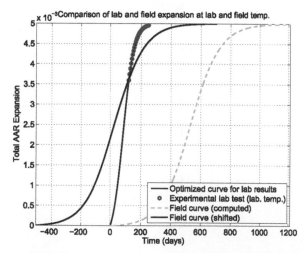

(e) Comparison of laboratory and field expansion at laboratory and field temperatures

FIGURE 9.7 Results

Sellier et al. (2009), in order to ultimately achieve a reasonable prediction of residual expansion. Further results will be presented in Saouma and Hariri-Ardebili (2013).

9.3   SUMMARY

Following a review of the (short) literature dealing with residual life, the outline of a new method is presented. Accounting for the thermodynamical nature of the reaction (i.e. time and temperature dependency), it assumes that the reaction can be indeed accelerated by temperature, and the existence of two coordinate systems. The first is the total expansion since concrete was cast, and the second is the residual accelerated expansion in the laboratory. A mathematical model is presented and solved through optimization techniques. The method appears to be promising, but much work remains to be performed.

References

ACI-207 (2005). *Guide to Mass Concrete (ACI 207.1R-05)*. American Concrete Institute.

ACI 318 (2011). *Building Code Requirements for Reinforced Concrete, (ACI 318-83)*. American Concrete Institute.

AFNOR P18-454 (2004). *AFNOR P18-454, Réactivité d'une formule de béton vis-à-vis de l'alcali-réaction (essaie de performance)*. Association Française de Normalisation, Paris.

Akhavan, A. (2001). Seismic safety evaluation of concrete arch dams. Technical report, M.Sc. Thesis in Hydraulic Structure Engineering; Sharif University of Technology (SUT).

Alvaredo, A. and Wittman, F. (1992). Crack formation due to hygral gradients. In *Third International Workshop on Behaviour of Concrete Elements Under Thermal and Hygral Gradients*. Weimar, Germany.

Anon. (1985). Design specifications of concrete arch dams. Technical report, Ministry of Water Conservancy and Electric Power of the People's Republic of China.

Arrhenius, S. (1889). On the reaction velocity of the inversion of cane sugar by acids. *Zeitschrift für physikalische Chemie*, 4:226.

ASTM (2012). *Concrete and Aggregates; Volume 04.02*.

ASTM C1260 (2000). *C1260-94 Standard Test Method for Potential Alkali Reactivity of Aggregates (Mortar-Bar Method)*. Annual Book of ASTM Standards.

ASTM C1293-08b (2008). *C1293-08b Standard Test Method for Determination of Length Change of Concrete Due to Alkali-Silica Reaction*. Annual Book of ASTM Standards.

ASTM C227 (2000). *C227-10 Standard Test Method for Potential Alkali Reactivity of Cement-Aggregate Combinations (Mortar-Bar Method)*. Annual Book of ASTM Standards.

ASTM C342-97 (1997). *ASTM C342-97 Standard Test Method for Potential Volume Change of Cement-Aggregate Combinations*. Annual Book of ASTM Standards. Withdrawn in 2001.

B., N.-M. M. (1992). *Mixed-Mode Fracture of Concrete: an Experimental Approach*. PhD thesis, Delft University of Technology.

Ballatore, E., Carpinteri, A., and Ferrara, G. (1990). Mixed mode fracture energy of concrete. *Engineering Fracture Mechanics*, 35(1–3):145–157.

Bangert, F., D., K., and Meschken, G. (2004). Chemo-hygro-mechanical modelling and numerical simulation of concrete deterioration caused by alkali–silica reaction. *Int. J. Numer. Anal. Meth. Geomech.*, 28:689–714.

Bažant, Z. and Becq-Giraudon, E. (2001). Estimation of fracture energy from basic characteristics of concrete. In de Borst, R., Mazars, J., Pijaudier-Cabot, G., and van Mier, J., editors, *Fracture Mechanics of Concrete Structures (Proc., FraMCoS-4 Int. Conf., Paris)*, pages 491–495. Balkema.

Bažant, Z. and Oh, B.-H. (1983). Crack band theory for fracture of concrete. *Materials and Structures*, 16:155–177.

Bažant, Z.P. (1984). Size effect in blunt fracture: Concrete, rock, metal. *J. of Engineering Mechanics, ASCE*, 110(4):518–535.

Bažant, Z.P. and Oh, B.H. (1983). Crack band theory for fracture of concrete. *Materials and Structures*, 93:155–177.

Bažant, Z.P and Steffens, A. (2000). Mathematical model for kinetics of alkali-silica reaction in concrete. *Cement and Concrete Research*, 30:419–428.

Bazant, Z. and Pfeifer, P. (1986). Shear fracture tests of concrete. *Materiaux et Constructions*, 19(110):111–121.

Ben Haha, M. (2006). *Mechanical effects of alkali silica reaction in concrete studied by SEM-image analysis*. PhD thesis, Ecole Polytechnique Federale de Lausanne. Thèse EPFL 3516.

Bérubé, M., Duchesnea, J., Doriona, J., and Rivestb, M. (2002). Laboratory assessment of alkali contribution by aggregates to concrete and application to concrete structures affected by alkali–silica reactivity. *Cement and Concrete Research*, 32:1215–1227.

Bérubé, M., Smaoui, N., Fournier, B., Bissonnette, B., and Durand, B. (2005). Evaluation of the expansion attained to date by concrete affected by alkali-silica reaction. part iii: Application to existing structures. *Canadian Journal of Civil Engineering*, 32(3):463–479.

Blight, G. and Alexander, M. G. (2008). *Alkali-Aggregate Reaction and Structural Damage to Concrete*. Taylor & Francis.

Bon, E., Chille, F., Masarati, P., and Massaro, C. (2001). Analysis of the effects induce by alkali-aggregate reaction (aar) on the structural behavior of Pian Telessio dam. In *Sixth International Benchmark Workshop on Numerical Analysis of Dams (Theme A)*, Salzburg, Austria.

Bournazel, J. and Moranville, M. (1997). Durability of concrete: The crossroad between chemistry and mechanics. *Cement and Concrete Research*, 27(10):1543–1552.

Brühwiler, E. and Wittmann, F. (1990). Failure of Dam Concrete Subjected to Seismic Loading Conditions. *Engineering Fracture Mechanics*, 35(3):565–572.

Brühwiler, E. (1988). *Fracture Mechanics of Dam Concrete Subjected to Quasi-Static and Seismic Loading Conditions*. Doctoral Thesis No 739, Swiss Federal Institute of Technology, Lausanne. (in German).

Capra, B. and Bournazel, J. (1998). Modeling of induced mechanical effects of alkali-aggregate reactions. *Cement and Concrete Research*, 28(2):251–260.

Capra, B. and Sellier, A. (2003). Orthotropic modeling of alkali-aggregate reaction in concrete structures: Numerical simulations. *Mechanics of Materials*, 35:817–830.

Carol, I. (1994). A unified theory of elastic degradation and damage based on a loading surface. *Int. J. Solids Structures*, 31(30):2835–2865.

Carol, I. and Bažant, Z.P. and Prat, P.C. (1992). Microplane type constitutive models for distributed damage and localized cracking in concrete structures. In *Proc.*

Fracture Mechanics of Concrete Structures, pages 299–304, Breckenridge, CO. Elsevier.

CEB (1983). Concrete under multiaxial states of stress constitutive equations for practical design. Technical Report Bulletin d'Information 156, Comité Euro-International du Béton.

Cedolin, L., Dei Poli, S., and Iori, I. (1987). Tensile behavior of concrete. *J. of Engineering Mechanics, ASCE*, 113(3).

Červenka, J. and Červenka, V. (1999). Three dimensional combined fracture-plastic material model for concrete. In *5th U.S. National Congress on Computational Mechanics*, Boulder, CO.

Cervenka, J., Chandra, J., and Saouma, V. (1998). Mixed mode fracture of cementitious bimaterial interfaces; part ii: Numerical simulation. *Engineering Fracture Mechanics*, 60(1):95–107.

Cervenka, J. and Papanikolaou, V. (2008). Three dimensional combined fracture-plastic material model for concrete. *International Journal of Plasticity*, 24(12):2192–2220.

Charlwood, R. (2012). Personal communication.

Charlwood, R. G., Steele, R., Solymar, S. V., and Curtis, D. D. (1992). A review of alkali aggregate reactions in hydroelectric plants and dams. In CEA and CAN-COLD, editors, *Proceedings of the International Conference of Alkali-Aggregate Reactions in Hydroelectric Plants and Dams*, pages 1–29, Fredericton, Canada.

Chatterji, S. (2005). Chemistry of alkali-silica reaction and testing of aggregates. *Cement & Concrete Composites*, pages 788–795.

Chiaramida, A. (2013). Lab to test seabrook's concrete problems. http://www.newburyportnews.com/local/x1862027937/Lab-to-test-Seabrooks-concrete-problems. Retrieved June 2013.

Comby-Peyrot, I., Bernard, F., Bouchard, P., Bay, F., and Garcia-Diaz, E. (2009). Development and validation of a 3d computational tool to describe concrete behaviour at mesoscale. application to the alkali-silica reaction. *Computational Materials Science*, 46(4):1163–1177.

Comi, C., Fedele, R., and Perego, U. (2009). A chemo-thermo-damage model for the analysis of concrete dams affected by alkali-silica reaction. *Mechanics of Materials*, 41(3):210 – 230.

Comi, C., Kirchmayr, B., and Pignatelli, R. (2012). Two-phase damage modeling of concrete affected by alkali–silica reaction under variable temperature and humidity conditions. *International Journal of Solids and Structures*, pages 3367–3380.

Comi, C. and Perego, U. (2011). Anisotropic damage model for concrete affected by alkali-aggregate reaction. *International Journal of Damage Mechanics*, 20(4):598–617.

Constantiner, D. and Diamond, S. (2003). Alkali release from feldspars into pore solutions. *Cement and Concrete Research*, 33:549–554.

Coppel, F., Lion, M., Vincent, C., and Roure, T. (2012). Approaches developed by edf with respect to the apprehension of risks of internal expansion of the concrete on nuclear structures: Management of operating power plants and prevention for new power plants. In *NUCPERF 2012 on Long-Term Performance of Cementitious*

Barriers and Reinforced Concrete in Nuclear Power Plant and Radioactive Waste Storage and Disposal, Cadarache, France.

Courant, R., Friedrichs, K., and Lewy, H. (1967). On the partial difference equations of mathematical physics. *IBM Journal of Research and Development*, 11(2):215–234.

CSA (2000). Guide to the evaluation and management of concrete structures affected by alkali -aggregate reaction.

Curtis, D. (1995). Modeling of aar affected structures using the grow3d fea program. In *Second International Conference on Alkali-Aggregate Reactions in Hydroelectric Plants and Dams*.

Dahlblom, O. and Ottosen, N. S. (1990). Smeared crack analysis using a generalized fictitious crack model. *Journal of Engineering Mechanics*, 116(1):55–76.

de Borst, R. (1986). *Nonlinear Analysis of Frictional Materials*. PhD thesis, Delft University of Technology, Delft.

de Borst, R. (2003). Preface. *Revue Francaise de Genie Civil*, 7(5):539–541.

de Borst, R. and Remmers, J.C. and Needleman, A. and Abellan, M.A. (1989). Discrete vs smeared crack models for concrete fracture: bridging the gap. *Eng. Computations*, 6:272–280.

Dennis, J. and Schnabel, R. (1983). *Numerical Methods for Unconstrained Optimization and Nonlinear Equations*. Englewood Cliffs.

Divet, L., Fasseu, P., Godart, B., Kretz, T., Laplaud, A., Le Mehaute, A., Le Mestre, G., Li, K., Mahut, B., Michel, M., Renaudin, F., Reynaud, P., Seignol, J., and Toutlemonde, F. (2003). Aide à la gestion des ouvrages atteints de réactions de gonflement interne: Guide méthodologique. Technical report, Laboratoire Central des Ponts et Chaussees (LCPC).

Dolen, T. (2005). Materials properties model of aging concrete. Technical Report Report DSO-05-05, U.S. Department of the Interior, Bureau of Reclamation, Materials Engineering and Research Laboratory, 86-68180.

Dolen, T. (2011). Selecting strength input parameters for structural analysis of aging concrete dams. In *21st Century Dam Design; Advances and Adaptations*, pages 227–243, San Diego, California. 31st Annual USSD Conference.

Dron, R. and Privot, F. (1992). Thermodynamic and kineic approach to the alkali-silica reaction. part i: Concepts. *Cement Concrete Research*, 22:941–948.

Duda, H. (1990). Stress-crack-opening relation and size effect in high-strength concrete. *Darmastad Concrete*, 5:35–46.

Dunant, C. and Scrivener, K. (2010). Micro-mechanical modelling of alkali-silica-reaction-induced degradation using the amie framework. *Cement and Concrete Research*, 40:517–525.

Espenson, J. (2002). *Chemical Kinetics and Reaction Mechanisms*. McGraw-Hill.

Fairbairn, E. M. R., Ribeiro, F. L. B., Lopes, L. E., Toledo-Filho, R. D., and Silvoso, M. M. (2006). Modeling the structural behavior of a dam affected by alkali-silica reaction. *Communications in Numerical Methods in Engineering*, 22:1–12.

Farage, M., Alves, J., and Fairbairn, E. (2004). Macroscopic model of concrete subjected to alkali–aggregate reaction. *Cement and Concrete Research*, 34:495–505.

Fasseu, P. (1997). *Alcali-réaction du béton: Essai d'expansion résiduelle sur béton durci*. Technical Report No. 44, LCPC. in French).

Feenstra, P. (1993). *Computational aspects of biaxial stress in plain and reinforced concrete. PhD Thesis*. PhD thesis, Delft University of Technology, The Netherlands.

Feenstra, P., de Borst, R., and Rots, J. (1991). Numerical study on crack dilatancy. i: Models and stability analysis. ii: Applications. *J. Eng. Mech.*, 117(4):733–769.

Fenwick, R. and Paulay, T. (1968). Mechanism of shear resistance of concrete beams. *J. of Struct. Division*, ASCE, 94(10):2325–2350.

FHWA (2006). Petrographic methods of examining hardened concrete: A petrographic manual. Technical Report FHWA-HRT-04-150, Federal Highway Administration.

FHWA (2010). Report on the diagnostis, prognosis, and mitigation of alkali-silica reaction (asr) in transportation structures. Technical Report FHWA-HIF-09-004, Federal Highway Administration.

FHWA (2013). FHWA ASR Reference Center. http://www.fhwa.dot.gov/pavement/concrete/asr/reference.cfm.

Fujiwara, K., Ito, M., Sasanuma, M., Tanaka, H., Hirotani, K., Onizawa, K., Suzuki, M., and Amezawa, H. (2009). Experimental study of the effect of radiation exposure to concrete. In *20th International Conference on Structural Mechanics in Reactor Technology (SMIRT 20)*, Espoo, Finland. SMIRT20-Division I, Paper 1891.

Furusawa, Y., Ohga, H., and Uomoto, T. (1994). An analytical study concerning prediction of concrete expansion due to alkali-silica reaction. In Malhotra, editor, *Proc. of 3rd Int. Conf. on Durability of Concrete*, pages 757–780, Nice, France. SP 145-40.

Galdieri, E. (2011). Personal Communication.

Gallitre, E. and Dauffer, D. (2010). Ageing management of french NPP civil work structures. In *International Workshop on Ageing-Management of Nuclear Power Plants and Waste Disposal Structures*, Toronto, Canada. AMP 2010 – EFC Event 334.

Gallitre, E., Dauffer, D., and Lion, M. (2010). La tenue du genie civil/vision 60 ans. In *SFEN ST7 : Séminaire Génie-civil et Nucléaire : de la Conception à l'Exploitation*, Toronto, Canada. in French.

Gao, X., Cyr, M., Multon, S., and Sellier, A. (2013). A comparison of methods for chemical assessment of reactive silica in concrete aggregates by selective dissolution. *Cement & Concrete Composites*, 37:82–94.

Gao, X., Multon, S., Cyr, M., and Sellier, A. (2012). Contribution to the requalification of alkali-silica reaction (asr) damaged structures: Assessment of the asr advancement in aggregates. In Drimalas, T., Ideker, J., and Fournier, B., editors, *14th International Conference on Alkali Aggregate Reaction in Concrete*, Austin, TX.

Gilks, P. and Curtis, D. (2003). Dealing with the effects of aar on the water retaining structures at mactaquac gs. In *Proceedings of the 21st Congress on Large Dams*, pages 681–703, Montreal, Canada.

Giuriani, E. and Rosati, G. P. (1986). Behaviour of concrete elements under tension after cracking. *Studi e Ricerche, Corso di Perfezionamento per le Costruzioni in Cemento Armato*, 8:65–82. (in Italian).

Glasser, D. and Kataoka, N. (1981). The chemistry of alkali-aggregate reaction. *Cement Concrete Research*, 11(1):1–9.

Glasser, F. (1992). Chemistry of the alkali-aggregate reaction. In Swamy, R., editor, *The Alkali-silica Reaction in Concrete*, pages 30–54. Van Nostrand Reinhold, New York.

Goodman, R.E. and Taylor, R.C. and Brekke, T.C. (1968). A Model for the Mechanics of Jointed Rocks. *J. of the Soil Mechanics and FOundations Division ASCE*, 94:637–659.

Gopalaratnam, V. S. and Shah, P. S. (1985). Softening response of plain concrete in direct tension. *ACI Journal*, 82:310–323.

Graves, H., Le Pape, Y., Naus, D., Rashid, J., Saouma, V., Sheikh, A., and Wall, J. (2013). Expanded materials degradation assessment (EMDA), volume 4: Aging of concrete. Technical Report NUREG/CR-ORNL/TM-2011/545, U.S. Nuclear Regulatroy Commission.

Grimal, E. (2007). *Caractérisation des effets du gonflement provoqué par la réaction alcali-silice sur le comportement mécanique d'une structure en béton*. PhD thesis, Université Paul Sabatier, Toulouse, France.

Grimal, E., Sellier, A., LePape, Y., and Bourdarot, E. (2008). Creep, shrinkage, and anisotropic damage in alkali-aggregate reaction swelling mechanism part I: A constitutive model. *ACI Materials Journal*, 105(3).

Grob, H. (1972). Schwelldruck am beipiel des belchentunnels. In *Int. Symp. Für Untertagbau*, page 99–119, Luzern.

Haberman, S. (2013). Seabrook station nuclear plant license advancing. http://www.seacoastonline.com/articles/20130515-NEWS-305150371. Retrieved June 2013.

Hassanzadeh, M. (1992). *Behavior of Fracture Process Zone in Concrete Influenced by Simultaneously Applied Normal and Shear Displacements*. PhD thesis, Lund Institute of Technology.

Hayward, D., Thompson, G., Charlwood, R., and Steele, R. (1988). Engineering and construction options for the management of slow/late alkali-aggregate reactive concrete. In *16th International Congress of Large Dams, Q62'C R33'*, San Francisco.

Herrador, M., Martínez-Abella, F., and Dopico, J. (2008). Experimental evaluation of expansive behavior of an old-aged asr-affected dam concrete: Methodology and application. *Materials and Structures*, 41(1):173–188.

Herrador, M., Martínez-Abella, F., and Hoyo Fernández-Gago, R. (2009). Mechanical behavior model for asr-affected dam concrete under service load: formulation and verification. *Materials and Structures*, 42(2):201–212.

Hillerborg, A., Modéer, M., and Petersson, P. (1976). Analysis of crack formation and crack growth in concrete by means of fracture mechanics and finite elements. *Cement and Concrete Research*, 6(6):773–782.

Hilsdorf, H., Kropp, J., and Koch, H. (1978). The effects of nuclear radiation on the mechanical properties of concrete. Technical Report SP55-10, American Concrete Institute.

Hordijk, D. (1991). *Local approach to Fatigue of Concrete*. PhD thesis, Delft University of Technology.

Hordijk, D., Reinhardt, H., and Cornelissen, H. (1989). Fracture mechanics parameters of concrete from uniaxial tensile tests as influenced by specimen length. In Shah, S. and Swartz, S., editors, *RILEM-SEM Int. Conference on Fracture Mechanics of Rock and Concrete*, New-York. Springer-Verlag.

Hsu, T., Slate, F., Sturman, G., and Winter, G. (1963). Microcracking of plain concrete and the shape of the stress-strain curve. *Journal of the American Concrete Institute*, 60:209–224.

Huang, M. and Pietruszczak, S. (1999). Alkali-silica reaction: modeling of thermo-mechanical effects. *Journal of Engineering Mechanics, ASCE*, 125(4):476–487.

Hydrofrac Inc. (2012). http://www.hydrofrac.com/deformation_gage.pdf. Date retrieved, Nov. 1, 2012.

Ichikawa, T. and Kimura, T. (2002). Effect of nuclear radiation on alkali-silica reaction. *Journal of Nuclear Science and Technology*, 44(10):1281–1284.

Ichikawa, T. and Koizumi, H. (2002). Possibility of radiation-induced degradation of concrete by alkali-silica reaction of aggregates. *Journal of Nuclear Science and Technology*, 39(8):880–884.

Ichikawa, T. and Miura, M. (2007). Modified model of alkali-silica reaction. *Cement and Concrete Research*, 37:1291–1297.

Ingraffea, A. (1977). *Discrete Fracture Propagation in Rocks: Laboratory Tests and Finite Element Analysis*. PhD thesis, University of Colorado.

Ingraffea, A. and Panthaki, M. (1985). Analysis of shear fracture tests of concrete beams. In Meyer, C. and Okamura, H., editors, *Finite Element Analysis of Concrete Structures*, pages 151–173. ASCE.

Institution of Structural Engineers (1992). Structural effects of alkali-silica reaction. technical guidance on the appraisal of existing structures. Technical report, Report of an ISE task group.

Jabarooti, M. and Golabtoonchi, I. (2003). Alkali-aggregate reactivity in south-east of iran. In *Proceedings of the 21st Congress on Large Dams*, pages 53–62, Montreal, Canada.

Jeang, F. L. and Hawkins, N. M. (1985). Nonlinear analysis of concrete fracture. Structures and mechanics report, Department of Civil Engineering, University of Washington, Seattle, WA.

Jin, W., Meyer, C., and Baxter, S. (2000). Glasscrete- concrete with glass aggregate. *ACI Materials Journal*, 97(2):208–213.

Keuser, W. and Walraven, J. (1989). Fracture of plain concrete under mixed mode conditions. In Shah, S. P., Swartz, S. E., and Barr, B., editors, *Fracture of Concrete and Rock*, pages 625–634, U. K. Elsevier Applied Science.

Kolmar, W. (1986). *Beschreibung der Kraftuebertragung ueber Risse in nichtlinearen Finite-Element-Berechnungen von Stahlbetontragwerken*. PhD thesis, Darmstadt.

Kupfer, B. and Gerstle, K. (1973). Behavior of concrete under biaxial stresses. *ASCE Journal of the Engineering Mechanics Division*, 99(4):853–866.

Larive, C. (1998a). Apports combinés de l'experimentation et de la modelisation à la comprehension de l'alcali reaction et de ses effects mecaniques. Technical report, Laboratoire Central des Ponts et Chausses (LCPC), 58 bd Lefebvre, Paris Cedex 15.

Larive, C. (1998b). *Apports Combinés de l'Experimentation et de la Modélisation à la Comprehension del'Alcali-Réaction et de ses Effets Mécaniques*. PhD thesis, Thése de Doctorat, Laboratoire Central des Ponts et Chaussées, Paris.

LCPC (1997). Détermination de l'indice de fissuration d'un parement de béton; Méthode d'essai LCPC N0. 47. Technical Report 47, Laboratoire Central des Ponts et Chaussées, Paris, France.

Leemann, A., Hammerschlag, J., and Thalmann, C. (2008). Inconsistencies between different accelerated test methods used to assess alkali-aggregate reactivity. In Broeckmans, M. and Wigum, B., editors, *13th International Conference on Alkali Aggregate Reaction in Concrete*, pages 944–953, Trondheim, Norway.

Leemann, A. and Merz, C. (2012). An attempt to validate the concrete performance test with the degree of aar-induced damage observed in concrete structures. In Drimalas, T., Ideker, J., and Fournier, B., editors, *14th International Conference on Alkali Aggregate Reaction in Concrete*, Austin, TX.

Leemann, A. and Merz, C. (2013). An attempt to validate the ultra-accelerated microbar and the concrete performance test with the degree of aar-induced damage observed in concrete structures. *Cement and Concrete Research*, 49:29–37.

Léger, P., Côte, P., and Tinawi, R. (1996). Finite element analysis of concrete swelling due to alkali-aggregate reactions in dams. *Computers & Structures*, 60(4):601–611.

Lemarchand, E., Dormieux, L., and Ulm, F. (2001). A micromechanical approach to the modeling of swelling due to alkali-silica reaction. In Ulm, F., Bazant, Z., and Wittmann, F., editors, *International conference on creep, shrinkage and durability mechanics of concrete and other quasi-brittle materials; Creep, shrinkage and durability mechanics of concrete and other quasi-brittle materials*, Austin (Texas), USA.

Li, K. and Coussy, O. (2002). Concrete ASR degradation: from material modeling to structure assessment. *J. Concrete Science and Engineering*, 4:35–46.

Li, K., Coussy, O., and Larive, C. (2004). *Modélisation chimico-mécanique du comportement des bétons affectés par la réaction d'alcali-silice Expertise numérique des ouvrages d'art déegradés*. Technical Report 43, *Laboratoire central des ponts et chaussées*, Paris.

Li, V. C. and Liang, E. (1986). Fracture processes in concrete and fiber reinforced cementitious composites. *Journal of Engineering Mechanics, ASCE*, 112(6):566–586.

Lotfi, H. (1992). *Finite Element Analysis of Fracture of Concrete and Masonry Structures*. PhD thesis, University of Colorado, Boulder.

Malla, S. and Wieland, M. (1999). Analysis of an arch-gravity dam with a horizontal crack. *Computers & Structures*, 72(1):267–278.

Malvern, L. (1969). *Introduction to the Mechanics of Continuous Medium.* Prentice-Hall.

Matest Inc. (2012). http://www.matest.com/products/concrete/flat-jacks-tests-on-brickworks-0.aspx. Retrieved: Oct. 2012.

MATLAB (2013). *version 7.10.0 (R2012a).* The MathWorks Inc., Natick, Massachusetts.

Menétrey, P. and Willam, K. (1995). Triaxial failure criterion for concrete and its generalization. *ACI Structural Journal,* 92(3):311–318.

METI (2001). Advanced research on dam seismic design. Technical report, Ministry of Economy, Trade and Industry; Agency for Natural Resources and Energy Civil Engineering Association of Power, Tokyo, Japan. *In Japanese.*

Mills-Bria, B., Nuss, L., Dixon, L., Powell, C., Harris, D., and O'Connell, D. (2006). State-of-practice for the nonlinear structural analysis of dams at the bureau of reclamation. Technical report, U.S. Dept. of the Interior, Bureau of Reclamation.

Mindess, S. and Young, F. (1981). *Concrete.* Prentice-Hall, Inc.

Mirzabozorg, H. (2013). Personal Communication.

ML121160349 (2012). Structural assessment of seabrook station. Prepared by Prof. O. Bayrak.

ML121160422 (2012). Impact of alkali silica reaction on seabrook concrete structure. Prepared by NextEra.

ML12160A374 (2012). Safety evaluation report with open items related to the license renewal of seabrook station.

ML12199A295 (2012). Presentation by nextera.

ML12199A300 (2012). Seabrook advisory committee on reactor safeguards presentation slides.

ML13151A328 (2013). Response to confirmatory action letter. Prepared by NextEra.

Mohammed, T., Hamada, H., and Yamaji, T. (2003). Alkali-silica reaction-induced strains over concrete surface and steel bars in concrete. *ACI Materials,* 100(2):133–142.

Multon, S. (2004). *Evaluation expérimentale et théorique des effets mécaniques de l'alcali-réaction sur des structures modèles.* PhD thesis, Université de Marne la Vallée, France. réalisée en partenariat LCPC-EDF. Reprise dans la collection études et recherches des Laboratoires des Ponts et Chaussées, série Ouvrages d'Art, OA46, 424 pages, 182 réf., résumé anglais, LCPC, octobre 2004.

Multon, S., Barin, F., Godart, B., and Toutlemonde, F. (2008). Estimation of the residual expansion of concrete affected by alkali silica reaction. *Journal of Materials in Civil Engineering,* 20:54–62.

Multon, S., Leclainche, G., Bourdarot, E., and Toutlemonde, F. (2004). Alkali-silica reaction in specimens under multi-axial mechanical stresses. In *Proceedings of CONSEC 4 (Concrete Under Severe Conditions),* pages 2004–2011, Seoul.

Multon, S., Sellier, A., and Cyr, M. (2009). Chemo-mechanical modeling for prediction of alkali silica reaction (asr) expansion. *Cement and Concrete Research,* 39:490–500.

Multon, S. and Toutlemonde, F. (2006). Effect of applied stresses on alkali-silica reaction induced expansions. *Cement and Concrete Research*, 36(5):912–920.

Murazumi, Y., Hosokawa, T., Matsumoto, N., Mitsugi, S., Takiguchi, K., and Masuda, Y. (2005a). Study of th einfluence of alkali-silica reaction on mecahnical properties of reinforced concrete members. In *18th International Conference on Structural Mechanics in Reactor Technology (SMIRT 18)*, pages 2043–2048, Beijing, China. SMIRT18-H03-3.

Murazumi, Y., Watanabe, Y., Matsumoto, N., Mitsugi, S., Takiguchi, K., and Masuda, Y. (2005b). Study of the influence of alkali-silica reaction on structural behavior of reinforced concrete members. In *18th International Conference on Structural Mechanics in Reactor Technology (SMIRT 18)*, pages 2036–2042, Beijing, China. SMIRT18-H03-2.

National Oceanic and Atmospheric Administration (2013). Satellite and Information Servis (NESDIS). http://www7.ncdc.noaa.gov/CDO/cdogetsubquery.cmd.

Naus, D. (2007). Primer on durability of nuclear power plant reinforced concrete structures - a review of pertinent factors. Technical Report NUREG/CR-6927 ORNL/TM-2006/529, U.S. Nuclear Regulatory Commission.

Nelson, K. and Bawendi, M. (2013). Thermodynamics and kinetics; MIT Open Courseware. http://ocw.mit.edu/courses/chemistry/5-60-thermodynamics-kinetics-spring-2008/lecture-notes/.

Ngo, D. and Scordelis, A. (1967). Finite element analysis of reinforced concrete beams. *Journal of the American Concrete Institute*, 64(3).

Nilson, A. (1968). Nonlinear analysis of reinforced concrete by the finite element method. *Journal of the American Concrete Institute*, 65(9).

NUREG-0980 (2013). Nuclear regulatory legislation; 112th congress, 2d session. Technical Report NUREG-0980, United States Nuclear Regulatory Commission.

Olivier, J. (1989). A consistent characteristic length for smeared cracking models. *International Journal for Numerical Methods in Engineering*, 28(2):461–474.

Orbovic, N. (2011). Personal communication. Canadian Nuclear Safety Commission.

Pan, J., Feng, Y., Jin, F., and Zhang, C. (2013). Numerical prediction of swelling in concrete arch dams affected by alkaliaggregate reaction. *European Journal of Environmental and Civil Engineering*, 17(4):231–247.

Paulay, T. and Loeber, P. (1987). Shear transfer by aggregate interlock. In *Shear in Reinforced Concrete*, pages 1–15, Detroit, Michigan. ACI. Special Publication SP-42.

Petersson, P. (1981). Crack growth and development of fracture zones in plain concrete and similar materials. Technical Report TVBM 1006, Lund Institute of Technology, Lund, Sweden.

Peyras, L., Royet, P., and Laleu, V. (2003). Functional modeling of dam performance loss; application to the alkali-aggregate reaction mechanism -chambon dam. In *Proceedings of the 21st Congress on Large Dams*, pages 853–872, Montreal, Canada.

Pian, J., Feng, Y., Wang, J., Sun, C., Zhang, C., and Owen, D. (2012). Modeling of alkali-silica reaction in concrete: a review. *Front. Struct. Civ. Eng.*, 6(1):1–18.

Poole, A. (1992). Introduction to alkali-aggregate reaction in concrete. In Swamy, R., editor, *The Alkali-silica Reaction in Concrete*, pages 1–28. Van Nostrand Reinhold, New York.

Portugese National Committee on Large Dams (2003). Ageing process and rehabilitation of pracana dam. In *Proceedings of the 21st Congress on Large Dams*, pages 121–138, Montreal, Canada.

Poyet, S., Sellier, A., Capra, B., Foray, G., Torrenti, J., Cognon, H., and Bourdarot, E. (2007). Chemical modelling of alkali silica reaction: Influence of the reactive aggregate size distribution. *Materials and Structures*, 40:229–239.

Poyet, S., Sellier, A., Capra, B., Thevenin-Foray, G., Torrenti, J.-M., Tournier-Cognon, H., and Bourdarot, E. (2006). Influence of water on alkali-silica reaction: Experimental study and numerical simulations. *Journal of Materials in Civil Engineering*, 18(4):588–596.

Puatatsananon, W. (2002). *Deterioration of Reinforced and Massive Concrete: A Multi-Physics Approach*. PhD thesis, University of Colorado, Boulder.

Puatatsananon, W. and Saouma, V. (2013). Chemo-mechanical micro model for alkali-silica reaction. *ACI Materials Journal*, 110:67–78.

Raphael, J. (1984). Tensile strength of concrete. *ACI Journal*, pages 158–165.

Rashid, Y. (1968). Analysis of prestressed concrete pressure vessels. *Nuclear Engineering and Design*, 7(4).

Richardson, L. (1911). The approximate arithmetical solution by finite differences of physical problems involving differential equations, with an application to the stresses in a masonry dam. *Philosophical Transactions of the Royal Society, Series A.*, 210:307–357.

Rots, J. G. and Blaauwendraad, J. (1989). Crack models for concrete: Discrete or smeared? fixed, multi-directional or rotating? *HERON*, 34(1):1512–1533.

Saouma, V. (1980). *Finite Element Analysis of Reinforced Concrete; a Fracture Mechanics Approach*. PhD thesis, Cornell University, Department of Structural Engineering.

Saouma, V. (2009). KumoNoSu, a 3d interactive graphics mesh generator for merlin; user's manual. http://civil.colorado.edu/~saouma/pdf/kumo.pdf.

Saouma, V. (2013a). Applications of the cohesive crack model to concrete, ceramics and polymers. In *Proceedigs of the 8th International Conference on the Fracture Mechanics of Concrete and Structures*, Toledo, Spain. Invited Pleanry Lecture.

Saouma, V. (2013b). A proposed aging management program for alkali aggregate reactions in a nuclear power plant. *Nuclear Engineering and Design*. in preparation.

Saouma, V., Broz, J., and Boggs, H. (1991a). In-situ Field Testing for Fracture Properties of Dam Concrete. *ASCE, Journal of Materials in Civil Engineering*, 3(3):219–234.

Saouma, V., Broz, J., Brühwiler, E., and Boggs, H. (1991b). Effect of aggregate and specimen size on fracture properties of dam concrete. *ASCE, Journal of Materials in Civil Engineering*, 3(3):204–218.

Saouma, V. and Hariri-Ardebili, M. (2013). Prediction of residual aar expansion. *ACI Materials Journal*. In preparation.

Saouma, V., Martin, R., and Hariri-Ardebili, M. (2013). A mathematical model for the kinetics of alkali-silica reaction. *ACI Materials Journal*. Under Review.

Saouma, V. and Perotti, L. (2004). Constitutive model for alkali aggregate reaction. Technical report, Report No. 2, Submitted to the Swiss Federal Agency of Water and Geology,, Bienne, Switzerland.

Saouma, V. and Perotti, L. (2006). Constitutive model for alkali aggregate reactions. *ACI Materials Journal*, 103(3):194–202.

Saouma, V., Perotti, L., and Shimpo, T. (2007). Stress analysis of concrete structures subjected to alkali-aggregate reactions. *ACI Materials Journal*, 104(5):532–541.

Saouma, V., Perotti, L., and Uchita, Y. (2005). Aar analysis of poglia dam with merlin. In *8th ICOLD Benchmark Workshop on Numerical Analysis of Dams*, Wuhan, China.

Saouma, V. and Sellier, A. (2010). Numerical benchmark for the finite element simulation of expansive concrete. http://civil.colorado.edu/saouma/AAR.

Saouma, V., Červenka, J., and Reich, R. (2010). Merlin finite element user's manual. http://civil.colorado.edu/~saouma/pdf/users.pdf.

Sausse, J. and Fabre, J. (2012). Diagnosis of dams affected by swellin reactions: Lessons learned from 150 monitored concrete dams in France. *Dam Engineering*, XXIII(1):5–17.

Scrivener (2003). Personal Communication.

Scrivener (2005). Personal Communication.

Seignol, J. (2010). Minutes of TCS 219-ACS, Team TC ACS-M, Modelling of Structures Meeting held in March, 12 2010.

Seignol, J. (2011). Practical guidance on predictive aar modelllin larive, li, & coussy model. Technical report, RILEM TCS 219 ACS. Internal Draft.

Seignol, J. and Godart, B. (2012). A collective effort to propose practical guidance on the use of numerical models to re-assess aar-affected structures. In Drimalas, T., Ideker, J., and Fournier, B., editors, *14th International Conference on Alkali Aggregate Reaction in Concrete*, Austin, TX.

Sellier, A., Bourdarot, E., Multon, S., Cyr, M., and Grimal, E. (2009). Combination of structural monitoring and laboratory tests for assessment of alkali aggregate reaction swelling: application to gate structure dam. *ACI Material Journal*, pages 281–290.

Sellier, A., Bournazel, J., and Mebarki, A. (1995). Une Modelisation de la reaction alcalis-granulat integrant une description des phenomenes aleatoires locaux. *Materials and Structures*, 28:373–383.

Serata Geomechanics (2005). In-situ stress/property measurements for quantitative design and construction. Technical report, Serata Geomechanics, Inc.

Shayan, A., Wark, R., and Moulds, A. (2000). Diagnosis of aar in canning dam, characterization of the affected concrete and rehabilitation of the structure. In *Proc. of 11th Int. Conf. on AAR*, pages 1383–1392, Quebec, Canada.

Shimizu, H., Asai, Y., Sekimoto, H., Sato, K. amd Oshima, R., Takiguchi, K., Masuda, Y., and Nishiguchi, I. (2005a). Investigaiton of safety margin for turbine generator foundation affected by alkali-silica reaction based on non-linear structural analysis.

In *18th International Conference on Structural Mechanics in Reactor Technology (SMIRT 18)*, pages 2049–2059, Beijing, China. SMIRT18-H03-4.

Shimizu, H., Watanabe, Y., Sekimoto, H., Oshima, R., Takiguchi, K., Masuda, Y., and Nishiguchi, I. (2005b). Study on material properties in order to apply for structural analysis of turbine generator foundation affected by alkali-silica reaction. In *18th International Conference on Structural Mechanics in Reactor Technology (SMIRT 18)*, pages 2055–2060, Beijing, China. SMIRT18-H03-5.

Shon, H. (2008). *PERFORMANCE-BASED APPROACH TO EVALUATE ALKALI-SILICA REACTION POTENTIAL OF AGGREGATE AND CONCRETE USING DILATOMETER METHOD*. Phd dissertation, Department of Civil Engineering, Texas A&M University.

Simo, J., Kennedy, J., and Govindjee, S. (1988). Non-smooth multisurface plasticity and viscoplasticity. loading/unloading conditions and numerical algorithms. *Int. J. Numer. Methods Eng.*, 26(10):2161–2185.

Sims, F., Rhodes, J., and Clough, R. (1964). Cracking in norfork dam. *J. of the American Concrete Institute*.

Slowik, V., Kishen, C., and Saouma, V. (1998). Mixed mode fracture of cementitious bimaterial interfaces; part I: experimental results. *Engineering Fracture Mechanics*, 60(1):83–94.

Smaoui, N., Bérubé, M., Fournier, B., Bissonnette, B., and Durand, B. (2004). Evaluation of the expansion attained to date by concrete affected by alkali-silica reaction. part I: Experimental study. *Canadian Journal of Civil Engineering*, 31(5):826–845.

Stankowski, T. (1990). *Numerical Simulation of Progressive Failure in Particle Composites*. PhD thesis, University of Colorado, Boulder.

Stanton, T. (1940). Expansion of concrete through reaction between cement and aggregate. *Proceedings of ASCE*, 66:1781–1811.

Struble, L. and Diamond, S. (1981). Swelling properties of synthetic alkali silica gels. *Journal of the American Ceramic Society*, 64(11):652–655.

Suwito, A., Jin, W., Meyer, C., and Xi, Y. (1998). Theoretical modeling on expansion and damage due to alkali-silica reaction. In *Proceedings of 12th Engineering Mechanics Conference*, pages 1175–1178, San Diego, CA.

Suwito, A., Jin, W., Xi, Y., and Meyer, C. (2002). A mathematical model for the pessimum effect of asr in concrete. *Concrete Science and Engineering, RILEM*, 4:23–34.

Swamy, R. and Al-Asali, M. (1988a). Engineering properties of concrete affected by alkali-silica. *ACI Materials Journal*, pages 367–374.

Swamy, R. and Al-Asali, M. (1988b). Influence of alkali-silica reaction on the engineering properties of concrete. In Dodson, V., editor, *Alkalies in Concrete, ASTM STP 930*, pages 69–86, Philadelphia, PA.

Swartz, S., Lu, L., and Tang, L. (1988). Mode ii fracture parameter estimates for concrete from beam specimens. *Experimental Mechanics*, 28(2):146–153.

Swartz, S. and Taha, N. (1990). Mixed mode crack propagation and fracture in concrete. *Engineering Fracture Mechanics*, 35:137–144.

Takagkura, T., Masuda, H., Murazumi, Y., Takiguchi, K., Masuda, Y., and Nishiguchi, I. (2005). Structural soundness for turbine-generator foundation affected by alkali-silica reaction and its maintenance plans. In *18th International Conference on Structural Mechanics in Reactor Technology (SMIRT 18)*, pages 2055–2060, Beijing, China. SMIRT18-H03-5.

Takatura, T., Ishikawa, T., Matsumoto, N., Mitsuki, S., Takiguchi, K., and Masuda, Y. (2005a). Investigaiton of the expanded value of tubine generator foundation affected by alkali-silica reaction. In *18th International Conference on Structural Mechanics in Reactor Technology (SMIRT 18)*, pages 2061–2068, Beijing, China. SMIRT18-H03-7.

Takatura, T., Watanabe, Y., Hosokawa, T., Ishii, T., Takiguchi, K., and Masuda, Y. (2005b). Vibration measurement and simulation analysis on a reinforced concrete structure with alkali-silica reaction. In *18th International Conference on Structural Mechanics in Reactor Technology (SMIRT 18)*, pages 2026–2035, Beijing, China. SMIRT18-H03-1.

Tcherner, J. and Aziz, T. (2009). Effects of aar on seismic assessment of nuclear power plants for life extensions. In *20th International Conference on Structural Mechanics in Reactor Technology (SMIRT 20)*, Espoo, Finland. SMIRT20-Division 7 Paper 1789.

Thompson, G., Charlwood, R., Steele, R., and Curtis, D. (1994). Mactaquac generating station intake and spillway remedial measures. In , editor, *Proceedings for the Eighteenth International Congress on Large Dams*, volume 1, Q-68, R.24, pages 347–368, Durban, South-Africa.

Ulm, F., Coussy, O., Kefei, L., and Larive, C. (2000). Thermo-chemo-mechanics of asr expansion in concrete structures. *ASCE J. of Engineering Mechanics*, 126(3):233–242.

U.S. Energy Information Administration (2013). Frequently asked questions. http://www.eia.gov/tools/faqs/faq.cfm?id=228&t=21. Date retrieved, Jun. 19, 2013.

USACE (1994). Arch dam design. http://www.usace.army.mil/publications/eng-manuals/em1110-2-2201/entire.pdf.

van Mier, J. (1984). Strain-softening of concrete under multiaxial loading conditions.

van Mier, J. (1986). Multi-axial strain softening of concrete; part i: Fracture. *Materials and Structures, RILEM*, 19(111).

Van Mier, J., Schlangen, E., and Nooru-Mohamed, n. (1992). Shear fracture in cementitious composites. In *Proc. Fracture Mechanics of Concrete Structures*, Breckenridge, CO. Elsevier.

Červenka, J., Kishen, C., and Saouma, V. (1998). Mixed mode fracture of cementitious bimaterial interfaces; part II: numerical simulation. *Engineering Fracture Mechanics*, 60(1):95–107.

Vodák, F. and Trtík, K. and Sopko, V. and Kapičková, O. and Demo, P. (2004). Effect of γ-irradiation on strength of concrete for nuclear safety-related structures. *Cement and Concrete Research*, 35:1447–1451.

Wagner, C. and Newell, V. (1995). A review of the history of aar at three of the tva's dam. In *Proc. of the Second International Conference on AAR in Hydroelectric Plants and Dams*, pages 57–66, Tennessee.

Walraven, J. C. (1980). *Discontinuous modeling of strain localisation and failure.* PhD thesis, Delft University of Technology, Delft, Netherlands. (in English).

West, G. (1996). *Alkali-Aggregate Reaction in Concrete Roads and Bridges.* Thomas Telford Ltd.

Wilkins, M. (1964). Calculation of elasto-plastic flow. *Methods of Computational Physics,* 3:211–263.

Wittke, W. and M., W. (2005). Design, construction and supervision of tunnels in swelling rock. In Erdem and Solak, editors, *Underground Space Use: Analysis of the Past and Lessons for the Future,* London. Taylor & Francis Group.

Wittmann, F., Rokugo, K., Brühwiler, E., Mihashi, H., and Simonin, P. (1988). Fracture Energy and Strain Softening of Concrete as Determined by Means of Compact Tension Specimens. *Materials and Structures,* 21:21–32.

Zienkiewicz, O. (1967). *The Finite Element in Structural and Continuum Mechanics.* McGraw-Hill, London, first edition.

Zienkiewicz, O. C., Taylor, R. L., and Zhu, J. (2005). *The Finite Element Method for Solid and Structural Mechanics.* Elsevier Butterworth-Heinemann, 6th edition.

A Numerical Benchmark for the Finite Element Simulation of Expansive Concrete

A.1 INTRODUCTION

A number of structures worldwide are known to (or will) suffer from chemically induced expansion of the concrete. This includes not only the traditional alkali aggregate reaction (also known as alkali silica reaction) but increasingly delayed ettringite formation (DEF)[1].

There are three components to the investigation of structures suffering from such an internal deterioration: a) Chemo-physical characterization focusing primarily on the material; b) Computational modeling of the evolution of damage and assessing the structural response of the structure; and c) managing the structure, (Divet et al., 2003).

Focusing on the second one, ultimately an engineer must make prediction for the response of a structure. In particular: a) is the structure operational, b) is it safe, and c) how those two criteria will evolve in time. This task is best addressed through a numerical simulation (typically finite element analysis) which should account for most of the structure's inherent complexities. This is precisely the object of this document.

The assessment of these finite element codes has been partially assessed within the ICOLD International Benchmark Workshops on Numerical Analysis of Dams[2], and there were only limited discussion of AAR within the European project *Integrity Assessment of Large Concrete Dams, NW-IALAD*[3] however there has not yet been any rigorous and rational assessment of codes.

Ultimately, practitioners would like to be able to calibrate their model with the limited historical field observation (typically inelastic crest displacements for dams, or crack maps for reinforced concrete) and then use it to extrapolate the behavior of the existing or modified structure into the future. In science and engineering, any extrapolation should be based on a fundamentally sound model which ideally should be independently assessed for its capabilities. Unfortunately, expansive concrete (finite element) models have not yet been assessed within a formal framework. The objective

---

[1]It is well known that DEF is often associated with AAR, however it is increasingly observed that it can occur by itself in massive concrete structure subjected to early age high temperature and under high relative humidity (above 95%).

[2]The sixth (Salzburg) and the eighth (Wuhan) benchmarks invited participants to analyze Pian Telessio and Poglia dams respectively. There was no submission to the former, and only two for the second.

[3] Survey of Current Calculation Practice.

of this effort is indeed an attempt to develop such a formal approach for the benefit of the profession.

Though we are aware of the importance of the chemical constituents of a reactive concrete (part a above), and their potential impact on the residual swelling, this aspect is not considered in this study. Henceforth, we limit ourselves to the interaction of various mechanical aspects: temperature, relative humidity, chemically induced swelling, and mechanical load.

The authors believe that prior to the comparison of analysis of a structures, a series of simple tests should first be undertaken. Each one of the test problems in turn will highlight a strength (or deficiency) of a model, one at a time. Then and only then, we could assess a model predictive capabilities for the analysis of a structure. This document will describe such a series of tests, and format in which data should be reported. In order to facilitate comparison, the test problems are of increasing complexity. For the most part we assess one parameter at a time, then two, and then three. Only after such an exercise could we compare full blown dam (or structural) analysis.

A.1.1 OBJECTIVES

This document is submitted by the authors to the Engineering community for the assessment of finite element codes which can perform a "modern" simulation of reactive concrete expansion. It will be posted under both http://dam.colorado.edu and http://www-lmdc.insa.fr/pres/modelgc, and submission could be made to either one.

The study is composed of two parts, the first addresses material modeling, and the second structure modeling. For the material modeling each study is split in two parts: a) parameter identification for the constitutive model (through calibration of your model with provided laboratory test results); and b) Prediction.

It is hoped that there will be sufficient interest and participation in this numerical benchmark study to eventually warrant a dedicated workshop.

A.1.2 IMPORTANT FACTORS IN REACTIVE CONCRETE

Assuming that the final residual swelling of the reactive concrete is known, and based on experimental and field observations, indications are that the following factors[4] should be considered in the finite element analysis of a structure:

1. Environmental Conditions of the concrete
   a. Temperature
   b. Humidity
2. Constitutive models
   a. Solid concrete (tension, compression, creep, shrinkage)
   b. Cracks/joints/interfaces.

---

[4]There is no general agreement on the importance of all these parameters, the list is intended to be inclusive of all those perceived by researchers to be worth examining.

3. Load history
4. Mechanical Boundary Conditions
    a. Structural Arrangement
    b. Reinforcement
    c. Anchorage

A.2 TEST PROBLEMS

A.2.1 P0: FINITE ELEMENT MODEL DESCRIPTION

Provide up to five pages of description of the model adopted in this particular order:

Constitutive Model

1. Basic principles of the model and its implementation.
2. Nonlinear constitutive model of sound or damaged concrete (clarify)
    a. Instantaneous response (elasticity, damage, plasticity, fracture and others)
    b. Delayed response (creep and shrinkage)
3. Effect on the chemically induced expansion by
    a. Moisture
    b. Temperature
    c. Stress confinement
4. Effect on the mechanical properties of concrete by
    a. Expansion
    b. Shrinkage and creep

Finite Element Code Features

1. Gap Element
2. Coupled hydro-thermo-mechanical
3. Others

A.2.2 MATERIALS

In light of the preceding list of factors influencing AAR, the following test problems are proposed. All results are to be entered in the accompanying spreadsheet and formatting instruction strictly complied with (to facilitate model comparison).

A.2.2.1 P1: Constitutive Models

At the heart of each code is the constitutive model of concrete. This problem will assess the code capabilities to capture the nonlinear response in both tension and compression.

It should be noted that in some codes, (Sellier et al., 2009) the constitutive model is tightly coupled (in parallel) with the AAR expansion one (modeled as an internal pressure), in other, (Saouma and Perotti, 2006) it is more loosely coupled (in series) with the AAR (modeled as an additional strain).

A.2.2.1.1   Constitutive Model Calibration

Perform a finite element analysis of a 16 by 32 cm concrete cylinder with f'_c, f'_t and E equal to 38.4 MPa, 3.5 MPa and 37.3 GPa respectively[5]. Traction is applied on the top surface, and a frictionless base is assumed. Make and state any appropriate assumption necessary, use the following imposed strain histogram:

$$0 \rightarrow 1.5\frac{f'_t}{E} \rightarrow 0 \rightarrow 3\frac{f'_t}{E} \rightarrow 1.5\varepsilon_c \rightarrow 0 \rightarrow 3\varepsilon_c \qquad (1.1)$$

where $\varepsilon_c = -0.002$. If needed, the fracture energy G_F in tension and compression are equal to $100 Nm/m^2$ and $10,000\ Nm/m^2$ respectively.

A.2.2.1.2   Prediction

Repeat the previous analysis following an AAR induced expansion of 0.5%, you may use the experimentally obtained degradation curve, by (Institution of Structural Engineers, 1992) and published by Capra and Sellier (2003), Fig. A.1

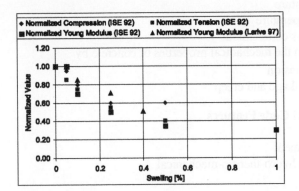

FIGURE A.1   Deterioration of AAR affected concrete (Capra and Sellier, 2003)

A.2.2.2   P2: Drying and Shrinkage

For some structures not necessarily under water (such as bridges or certain hydraulic structures), drying shrinkage strains may be of similar order of magnitude as the AAR induced ones. As shown in Fig. A.2 one must consider various cases of drying and shrinkage, reactive and non reactive concrete, and at relative humidities ranging from

---

[5]These parameters should be used in all subsequent test problems.

a low 30% to a fully saturated environment, and sealed or not. There are a total of 6 potential cases of interest:

a. Non reactive concrete at 30% RH
b. Reactive concrete at 30% humidity
c. Non Reactive concrete sealed specimen
d. Non Reactive concrete under water.
e. Reactive Concrete, sealed cylinder.
f. Reactive concrete under water.

which will be analyzed in P2 and P5

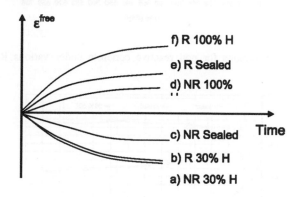

FIGURE A.2   Drying and Shrinkage test Cases

A.2.2.2.1   Constitutive Model Calibration

Fit your parameters using a 16 by 32 cm cylinder by performing the following analyses: a, c, and d with respect to the temporal variation of mass (Fig. A.3) and longitudinal strain (Fig. A.4)

Mass, Fig. A.3 and strain, Fig. A.4 temproal variation[6].

A.2.2.2.2   Prediction

Using the parameter determined from the previous section, repeat the same analysis with the temporal variation of external RH for the cylinder shown in Fig. A.5.

$$RH(\text{week}) = \frac{RH_{\text{max}} - RH_{\text{min}}}{2} \sin\left(2\pi \frac{t-16}{52}\right) + \frac{RH_{\text{max}} - RH_{\text{min}}}{2} \quad (1.2)$$

where RH_{max} and RH_{min} are equal to 95% and 60% respectively.

---

[6]All available experimental results are tabulated in separate Excel files.

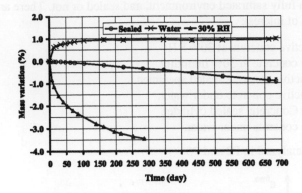

FIGURE A.3   Mass variations for non reactive concrete under various RH conditions;
(Multon, 2004)

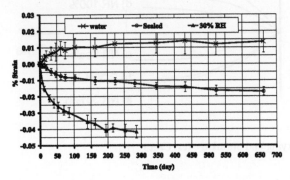

FIGURE A.4   Strain variations for non reactive concrete under various RH conditions; (Multon, 2004)

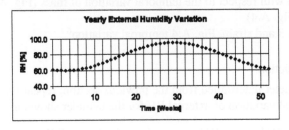

FIGURE A.5   Humidity variation

A.2.2.3   P3: Basic Creep

There is strong experimental and field indications that creep plays a dominant role
in the irreversible long term deformation concrete subjected to constant load. Its

effect must be accounted for to properly extract the AAR expansion. This may be explained through biaxially or triaxially loaded elements where swelling is restricted in one direction while free to occur on the other(s). Therefore, in the AAR constrained direction creep deformation will be predominant. This is more likely to occur in arch dams.

A.2.2.3.1 Constitutive Model Calibration

For a 13 by 24 cm cylinder subjected to 10 and 20 MPa axial compression, plot the longitudinal and radial displacements. You may calibrate your model on the experimental curve shown in Fig. A.6.

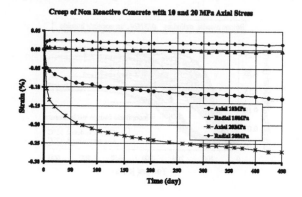

FIGURE A.6   Creep in non-reactive concrete under sealed condition for different axial stress; (Multon, 2004)

A.2.2.3.2 Prediction

Using the previously determined parameters, repeat the same analysis for the axial load history shown in Fig. A.7.

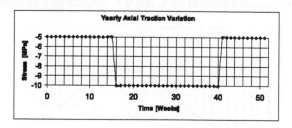

FIGURE A.7   Stress variation

A.2.2.4   P4: AAR Expansion; Temperature Effect

All chemical reactions are thermodynamically driven. Reactive concrete expansion varies widely with temperature ranges usually encountered in the field or laboratories. Hence, it is of paramount importance that the kinetics of the reaction captures this dependency.

A.2.2.4.1   Constitutive Model Calibration

Perform the finite element analysis of a 13 by 24 cm cylinder under water, free to deform at the base and undergoing a free expansion, and for $T = 23°C$ and $38°C$. Fit the appropriate parameters of your model with Fig. A.8 obtained by Larive (1998a).

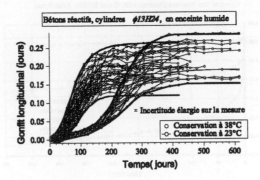

FIGURE A.8   Free expansion from Larive's tests; (Larive, 1998a)

A.2.2.4.2   Prediction

Repeat the previous analysis using the variable internal temperature

$$T(\text{week}) = \frac{T_{\max} - T_{\min}}{2} \sin\left(2\pi \frac{t - 16}{52}\right) + \frac{T_{\max} - T_{\min}}{2} \tag{1.3}$$

where $T_{\max}$ and $T_{\min}$ are equal to 25°C and 0°C respectively, as shown in Fig. A.9

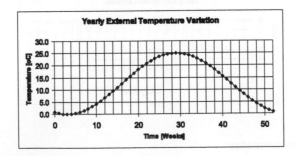

FIGURE A.9   Temperature variation

A.2.2.5 P5: Free AAR Expansion; Effect of RH

Relative humidity plays a critical role in the expansion of AAR affected concrete. It is well established, (Poole, 1992) that expansion will start for a RH at least equal to 80%, and will then increase with RH (RH^8 is a widely accepted forumula). For external bridge structures and some dams this can be critical.

A.2.2.5.1 Constitutive Model Calibration

Using a 16 by 32 cm cylinder, and assuming a temperature of 38oC, fit the appropriate parameters for mass and vertical strain variation of reactive concrete as shown in Fig. A.10 and A.11 respectively.

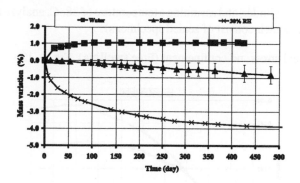

FIGURE A.10   Mass variation for reactive concrete under various RH conditions; (Multon, 2004)

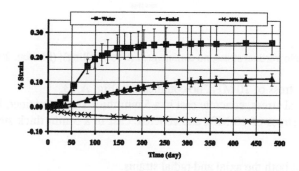

FIGURE A.11   Strain variation for reactive concrete under various RH conditions; (Multon, 2004)

A.2.2.5.2 Prediction

Repeat previous analysis using the RH variation shown in Fig. A.5.

A.2.2.6 P6: AAR Expansion; Effect of Confinement

It has long been recognized that confinement inhibits reactive concrete expansion, (Charlwood et al., 1992), (Léger et al., 1996) and most recently (Multon and Toutlemonde, 2006). This test series seeks to ensure that this is properly captured by the numerical model.

A.2.2.6.1 Constitutive Model Calibration

For a 13 by 24 cm cylinder, and assuming a temperature of 38oC, analyze the following test cases (all of which consist of sealed specimens):

P6-a. No vertical stress, no confinement (Free swelling), Fig. A.12.

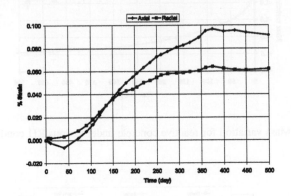

FIGURE A.12   No vertical stress, no confinement (free swelling); (Multon, 2004)

P6-b. Vertical stress of 10 MPa, no confinement, Fig. A.13.

P6-c. No vertical stress, concrete cast in a 5 mm thick steel container, Fig. A.14.

P6-d. Vertical stress of 10 MPa and concrete cast in a 5 mm thick steel container, Fig. A.15.

In all cases, plot both the axial and radial strains.

A.2.2.6.2 Prediction

Repeat the analysis with the vertical and radial imposed stress histogram shown in Fig. A.7.

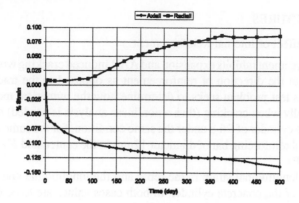

FIGURE A.13   10 MPa vertical stress, no confinement; (Multon, 2004)

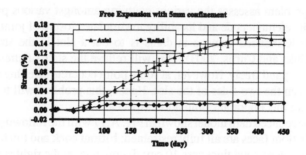

FIGURE A.14   Vertical stress of 10 MPa and concrete cast in a 5 mm thick steel container; (Multon, 2004)

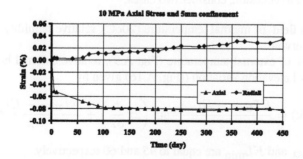

FIGURE A.15   Vertical stress of 10 MPa and concrete cast in a 5 mm steel container; (Multon, 2004)

A.2.3 STRUCTURES

A.2.3.1 P7: Effect of Internal Reinforcement

Internal reinforcement inhibits expansion and AAR induced cracking would then align themselves with the direction of reinforcement as opposed to the traditional "map cracking". This test problem seeks to determine how the numerical model accounts for this, specially when cracking (thus a nonlinear analysis is needed) occurs.

Analyze the cylinder of P6-a under the same condition, for the same duration with a single internal reinforcing bar of diameter 12mm in the center, and E=200,000 MPa and f_y=500MPa.

Determine longitudinal strain in the rebar and the longitudinal and radial strains on the surface of the concrete cylinder. In both cases values are to be determined at midheight.

A.2.3.2 P8: AAR Expansion; Idealized Dam

This last test problem assesses the various coupling amongst various parameters as well as the finite element code and its ability to simulate closure of joint. A common remedy for AAR induced damage in dams is to cut a slot in the structure as in Mactaquac, (Gilks and Curtis, 2003). This will relieve the state of stress, and allow the concrete to expand freely. However, at some point concrete swelling will result in a contact between the two sides of the slot. Hence, this problem will test the model ability to capture this important simulation aspect as well.

Consider the reduced dam model shown in Fig. A.17 with the following conditions: a) lateral and bottom faces are all fully restrained; b)front back and top faces are free; c) slot cut at time zero, total thickness 10 cm; d) concrete on the right is reactive, and concrete block on the left is not reactive; e) hydrostatic pressure is applied only on the right block.

Using the fitting data of P6, and an friction angle of $50°$ for concrete against concrete, and zero cohesion, consider two cases:

Homogeneous field of internal temperature (20oC), relative humidity (100%), and an empty reservoir.

Transient field of external temperature (Fig. A.9), relative external humidity (Fig. A.5), and pool elevation variation ((Fig. A.16) given by

$$EL(week) = \frac{EL_{max} - EL_{min}}{2} \sin\left(2\pi\frac{t}{52}\right) + \frac{EL_{max} - EL_{min}}{2} \quad (1.4)$$

where EL_{max} and EL_{min} are equal to 95 and 60 respectively.

For both analysis, the specified temperature and relative humidity is the one of the concrete surface. Zero flux condition between dam and foundation. Reference base temperature of the dam is 20°C.

1. x, y, z displacements of point A.

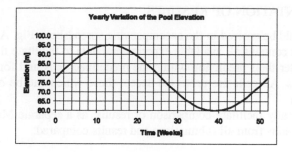

FIGURE A.16   Yearly variation of pool elevation

2. F_x, F_y and F_z resultant forces on the fixed lateral face versus time (25 years).
    Assume the typical yearly variations of external air temperature and pool
    elevation shown in Fig. A.9 and A.16 respectively.

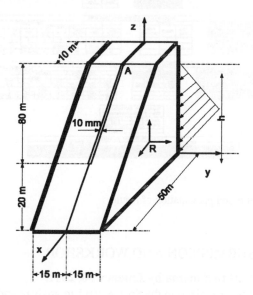

FIGURE A.17   Idealized dam

This model seeks to capture: a) general finite element program capabilities in modeling
the joint response; b) ease (or difficulty in preparing the input data file for a realistic
problem; and c) coupling of the various parameters.

A.3 PRESENTATION OF RESULTS

All results should be entered in the accompanying spreadsheet, Fig. A.18. Note that the spreadsheet contains all available experimental data to facilitate fits, and participants must enter their prediction within the predefined cells and for the specified time increments. All cells are protected except those which can be overwritten by participant data.

This will greatly facilitate comparison of results, as a separate Matlab program could extract results from all submissions and results compared.

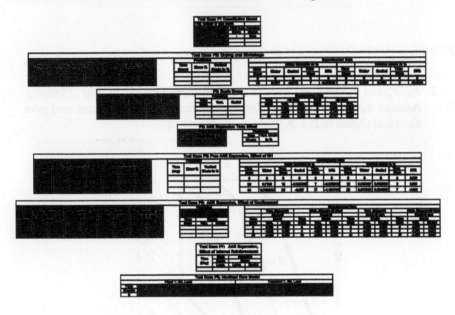

FIGURE A.18   Excel based presentation of results

A.4 RESULTS SUBMISSION AND WORKSHOP

Results should be sent to Saouma by January 30, 2011.

A workshop will be organized in the Spring 2011 to analyze and interpret results. The objectives being:

1. What would be the simplest model (irrespective of the code) which can reliably predict the long term behavior and expansion of concrete suffering from chemically induced expansion.
2. Determine which codes can model some/all aspects of the concrete expansion.
3. What are the important factors which should be accounted for in a numerical prediction.

There will be two lead reviewers to present, assess results

- Alain Carrere, Chairman of the ICOLD committee on Numerical Methods, formerly with Coyne et Bellier
- Prof. Jan Rots, from the Delft University of Technology, a leading authority in the finite element simulation of reinforced concrete.

Please check http://dam-colorado.edu/AAR for updates (and registration if you so desire).

We are seeking funding to (partially) support this endeavor.

A.5 ACKNOWLEDGEMENTS

Preparation of this Benchmark was made possible through the financial support of the Université Paul Sabattier of Toulouse which invited V. Saouma as a *Professeur Invité* in 2009, and the valuable assistance of Stephane Multon who performed some of the reported tests.

The authors thank Catherine Larive, and Stephane Multon who performed the reported and published tests in the *Laboratoire Central des Ponts et Chaussées* under the supervision of Olivier Coussy and Francois Toutlemonde respectively. These tests were made possible through the financial support of EdF-CIH (Eric Bourdarot).

B Merlin

B.1 INTRODUCTION

Complex nonlinear time-dependent analyses require special features not often found in commercial codes. The constitutive model developed in Chapter 2 at the very least requires an evaluation of the AAR-induced strain at each Gauss point (Eq. 2.1). From a practical point of view however and as shown in the referenced analyses, the following features must also be incorporated:

1. Transient (linear) thermal analysis
2. Transfer of nodal temperature from the thermal analysis to the stress analysis (different meshes are typically used due to the presence of interface elements)
3. Nonlinear joint elements for dam analysis and/or nonlinear continuum elements for reinforced concrete structures
4. Ability to reset displacements to zero while maintaining the stress-strain state (to separate deformations due to dead load from other loads)
5. Generation of a fairly complex input data file that specifies for each time increment: pool elevation, and nodal temperatures.

Data preparation and results interpretation should be conducted through "user-friendly" preprocessor / post-processor set-up.

This chapter will outline the tools necessary for a 3D AAR simulation using the Merlin finite element code (Saouma et al., 2010). All of these tools have been developed by the author and all run under Windows (as well as Unix).

B.2 ARCH DAM PREPROCESSOR: BEAVER

Though arch dams are defined by parametric curves that have been mathematically defined, we will require a numerical discretization in 3D. This step can be achieved using a tool such as Beaver, which is based on a mathematical definition of the dam, as described in MATLAB (2013), that subsequently generates patches in 3D. While a structural mesh can be derived for the dam, its base and foundation would contain unstructured meshes (Fig. B.1). A boundary representation will also be prepared for the formal mesh generator KumoNoSu described below.

B.3 PREPROCESSOR: KUMONOSU

The mesh generator, KumoNoSu[1] is a graphical front end to two programs (both written by Dr. Daniel Rypl):

---

[1] In Japanese: Spider web.

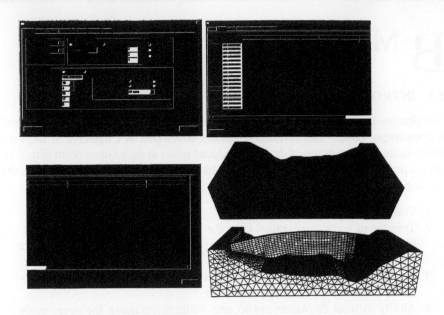

FIGURE B.1   Preliminary mesh generation for dams

T3D a powerful mesh generator capable of producing both unstructured (based on Delaunay triangularization) and structured meshes, citeprypl

T3D2Merlin which enables defining material properties, boundary conditions and loads for a Merlin input file.

Hence, KumoNoSu starts by producing a boundary definition of the physical object to be discretized, i.e. a .bd file. Once the boundaries of the solid have been delineated and mesh-size generation guidelines established for the code, the program constructs a mesh to describe the structure: this constitutes the .t3d file. Next, material properties and loading data must be added for each element to the geometric data set. This step is saved in a .ctrl file. In addition, a finite element analysis algorithm is available to process the expanded data set into the Merlin .inp file (Fig. B.2).

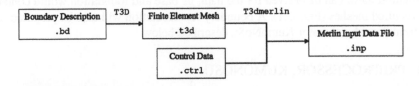

FIGURE B.2   KumoNoSu's File Types

The boundary representation for T3D serves as a geometric description of individual model entities and moreover reveals their topological relationships (Fig. B.3):

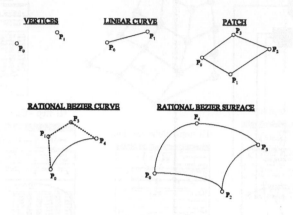

FIGURE B.3   Entities recognized by KumoNoSu

Vertices points in (x,y,z) space
Curves defined by 2 end vertices, may be linear, quadratic or cubic
Patches planar collection of curves
Surfaces non-planar, defined by (4) curves
Shells non-planar collection of curves
Regions set of non-self-intersecting boundary surfaces, patches and shells.

These entities are defined hierarchically (see Fig. B.4): Lower-level entities that be-

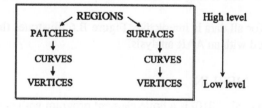

FIGURE B.4   Hierarchy of Model Representation

long to higher-level ones are called the **Minor** entities of the given higher-level entity.
Hence:

* Vertices are the minor entities of curves.
* Vertices and curves are the minor entities of patches and surfaces.
* Vertices, curves, surfaces and patches are the minor entities of regions.

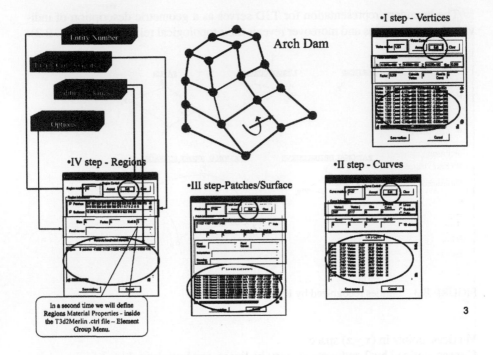

FIGURE B.5   Mesh definition process in KumoNoSu

Figure B.5 illustrates the mesh definition process.

It should be noted that material properties, loads and boundary conditions have not been defined with respect to the mesh, but rather the mesh boundary representation. Such a set-up allows fine-tuning the mesh density without the need to adjust these parameters.

The input step for all data is interactive; Figure B.6 illustrates the graphical user interfaces associated with an AAR analysis.

B.4   ANALYSIS: MERLIN

MERLIN (Saouma et al., 2010) is a finite element program with capabilities in both two and three dimensions:

- Static and dynamic (implicit and explicit) stress analyses
- Steady-state and transient heat (and seepage) analyses

The main features of the Merlin program are:

- Full support of the AAR model described in this book
- A rich library of elements, interface/joint elements (5) and constitutive models (20) for nonlinear steel, concrete, rock and joints

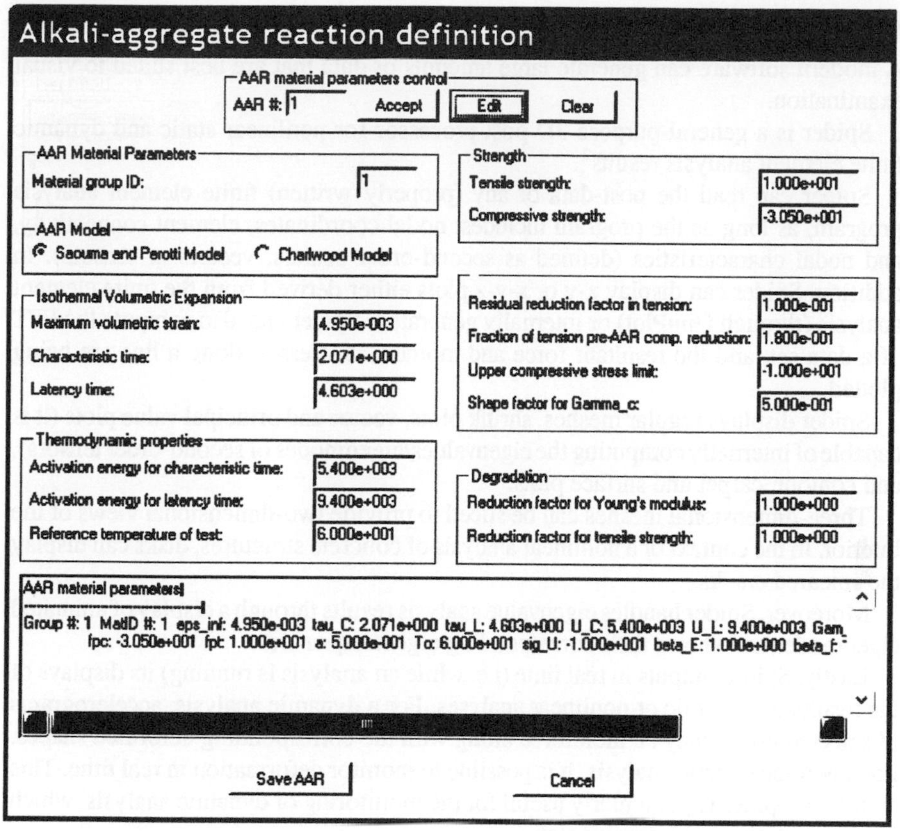

Alkali-aggregate reaction definition

AAR material parameters control
AAR #: |1| Accept | Edit | Clear

AAR Material Parameters
Material group ID: |1|

Strength
Tensile strength: |1.000e+001|
Compressive strength: |-3.050e+001|

AAR Model
(•) Saouma and Perotti Model   ( ) Charlwood Model

Isothermal Volumetric Expansion
Maximum volumetric strain: |4.950e-003|
Characteristic time: |2.071e+000|
Latency time: |4.603e+000|

Residual reduction factor in tension: |1.000e-001|
Fraction of tension pre-AAR comp. reduction: |1.800e-001|
Upper compressive stress limit: |-1.000e+001|
Shape factor for Gamma_c: |5.000e-001|

Thermodynamic properties
Activation energy for characteristic time: |5.400e+003|
Activation energy for latency time: |9.400e+003|
Reference temperature of test: |6.000e+001|

Degradation
Reduction factor for Young's modulus: |1.000e+000|
Reduction factor for tensile strength: |1.000e+000|

AAR material parameters|

Group #: 1 MatID #: 1 eps_inf: 4.950e-003 tau_C: 2.071e+000 tau_L: 4.603e+000 U_C: 5.400e+003 U_L: 9.400e+003 Gam_
fpc: -3.050e+001 fpt: 1.000e+001 a: 5.000e-001 To: 6.000e+001 sig_U: -1.000e+001 beta_E: 1.000e+000 beta_f:

Save AAR   Cancel

FIGURE B.6   Graphical user interface for inputting AAR properties in KumoNoSu

- Multiple integration schemes (Newton-Raphson, modified Newton-Raphson, secant, arc-length)
- Numerous features for the dynamic analysis of massive or reinforced concrete structures (fluid-structure, soil-structure, fluid-fracture interactions, restart, implicit/explicit time marching schemes)
- A special feature for dams: automatic uplift update to reflect crack propagation, gallery identification and drain effectiveness, mass foundation, definition, added mass or fluid elements, and automated 2D or 3D deconvolution
- Seamless integration with the KumoNoSu preprocessor and Spider postprocessor.

This code has been continuously developed / improved for over 20 years.

B.5   POST-PROCESSOR: SPIDER

A modern software can generate large amounts of data that are best suited to visual examination.

Spider is a general-purpose 3D post-processor for nonlinear static and dynamic finite element analysis results[2].

Spider can read the post-data of any (properly written) finite element analysis program, as long as the program includes: nodal coordinates, element connectivity, and nodal characteristics (defined as second-order scalars, vectors or tensors). In addition, Spider can display x-y or x-y-z plots either derived from the finite element analysis (through GnuPlot) or internally generated. Spider can also compute the FFT of a data set, and the resultant force and moment if stresses along a line are being plotted.

Spider displays regular meshes, shrink plots, vector and principal value plots (it is capable of internally computing the eigenvalues/eigenmodes of second-order tensors), and contour, carpet and surface plots.

Three-dimensional meshes can be sliced to provide two-dimensional views of the interior. In the context of a nonlinear analysis of concrete structures, disks can display the smeared cracks.

Moreover, Spider handles eigenvalue analysis results through a display of animated eigenmodes and shows their corresponding eigenfrequencies.

Lastly, Spider outputs in real time (i.e. while an analysis is running) its displays of the results of dynamic or nonlinear analyses. For a dynamic analysis, accelerograms of selected nodes may be monitored along with the corresponding deformed shapes. For a nonlinear static analysis, it is possible to monitor deformation in real time. This feature of Spider is particularly useful for the monitoring of dynamic analysis, which is a computationally intensive step.

Spider offers a mouse-oriented, graphical user interface, making the program easy and intuitive to run. No command prompts are introduced and no directives need to be memorized.

Spider input files are relatively straightforward to define and can be read in either binary or ASCII format. Since Spider understands various data types and reads the labels contained in the menus along with the post-data from the post file, the specific type of analysis performed by a finite element code does not affect Spider's programming. Menus will be visualized with proper labels and plots will present the data in the formats described below. Hence, Spider is not strictly limited (or tied) to stress analysis.

B.5.1   INTEGRATION

The seamless integration of the three codes is illustrated in Figure B.8, which reveals that practically no file handling is required.

---

[2]Spider was initially developed as the post-processor for the MERLIN program but has since been expanded for use with any finite element analysis program.

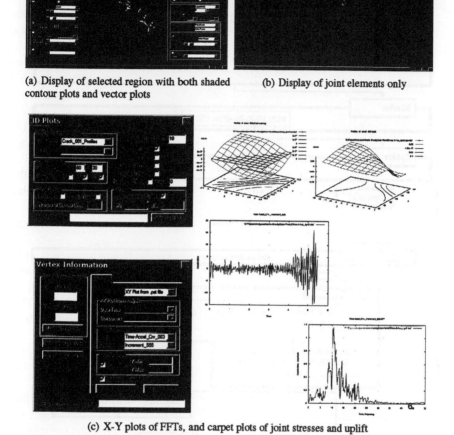

(a) Display of selected region with both shaded contour plots and vector plots

(b) Display of joint elements only

(c) X-Y plots of FFTs, and carpet plots of joint stresses and uplift

FIGURE B.7   Examples of Spider output

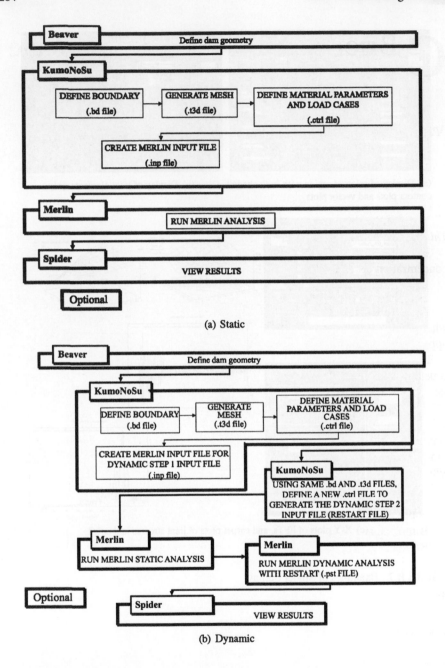

(a) Static

(b) Dynamic

FIGURE B.8   Integration of the KumoNoSu, Merlin and Spider codes

C Brief Review of Reaction Rate

Appendix written in collaboration with Mohammad Amin Hariri-Ardebili

C.1 DEFINITIONS

This section has been largely inspired by Nelson and Bawendi (2013).

The rate law or rate equation of a chemical reaction links the reaction rate k to the concentration of reactant $[\bullet]$.

$$aA + bB \rightarrow cC + dD \tag{3.1}$$

where lower case letters (a, b, c, and d) represent stoichiometric coefficients[1], while upper case letters denote the reactants (A and B) and products (C and D).

The reaction rate k for a chemical reaction occurring in a closed system under constant-volume conditions, without any build-up of reactive intermediates, is defined as:

$$r = -\frac{1}{a}\frac{d[A]}{dt} = -\frac{1}{b}\frac{d[B]}{dt} = \frac{1}{c}\frac{d[C]}{dt} = \frac{1}{d}\frac{d[D]}{dt} \tag{3.2}$$

where $[X]$ is the concentration of substance X in moles per volume of solution. It should be noted that the reaction rate is always positive and the $-$ sign is associated with the decreasing reactant concentration.

The reaction order can be classified as:

Unimolecular	$A \rightarrow$ products	1st order	$r = k(T)[A]$
Bimolecular	$A \rightarrow$ products	2nd order	$r = k(T)[A]^2$
Bimolecular	$A + B \rightarrow$ products	2nd order	$r = k(T)[A][B]$
Termolecular	$A + B + C \rightarrow$ products	3^{ed} order	$r = k(T)[A][B][C]$

$$\tag{3.3}$$

and

$$k(t) = Ae^{-\frac{E_0}{RT}} \tag{3.4}$$

is the Arrhenius Law previously used in Eq. 8.18.

---

[1]Stoichiometry is simply the mathematics applied in the field of chemistry. Given enough information, one can use stoichiometry to calculate masses, moles and percentages within a chemical equation.

C.2 EXAMPLES OF SIMPLE REACTIONS

C.2.1 ZERO-ORDER REACTIONS

A zero-order reaction has a rate independent of reactant concentration. Increasing the concentration of reacting species will not accelerate the reaction rate, i.e. the amount of substance reacted is proportional to the reaction time. For a simple reaction, such as $A \rightarrow$ products, the reaction rate is defined as:

$$r = -\frac{d[A]}{dt} = k \tag{3.5}$$

In solving this simple differential equation, we obtain the so-called integrated zero-order rate law:

$$[A]_t = -kt + [A]_0 \tag{3.6}$$

where $[A]_t$ is the concentration of the targeted chemical at a specific time, $[A]_0$ represents the initial concentration, and k is expressed in [moles/(liter.sec)].

A reaction is of a zero order if the concentration data plotted versus time yield a straight line. The slope of this resulting line is the negative of the zero-order rate constant k, as shown in Figure C.1(a). The half-life (i.e. time needed for half the reactant to be depleted) of this equation is given by:

$$t_{\frac{1}{2}} = \frac{[A]_0}{2k} \tag{3.7}$$

C.2.2 FIRST-ORDER REACTIONS

A first-order reaction depends on the concentration of just one reactant. Other reactants may be present, but each will be of a zero order. For a simple reaction, such $A \rightarrow$ $products$, the reaction rate is defined as:

$$r = -\frac{d[A]}{dt} = k[A] \tag{3.8}$$

where k is in [1/sec], and the integrated first-order rate law is:

$$ln[A]_t = -kt + ln[A]_0 \text{ or } [A]_t = [A_0]e^{-kt} \tag{3.9}$$

A plot of $ln[A]_t$ vs. time t yields a straight line with a slope of $-k$ (see Fig. C.1(b)). The half-life (time needed for half the reactant to be depleted) of this equation is then given by:

$$t_{\frac{1}{2}} = \frac{\ln 2}{k} \tag{3.10}$$

C.2.3 SECOND-ORDER REACTIONS

A second-order reaction depends on the concentrations of a second-order reactant or else two first-order reactants. For the former, such as $A \rightarrow$ products, the reaction rate is defined as:

$$-\frac{d[A]}{dt} = k[A]^2 \tag{3.11}$$

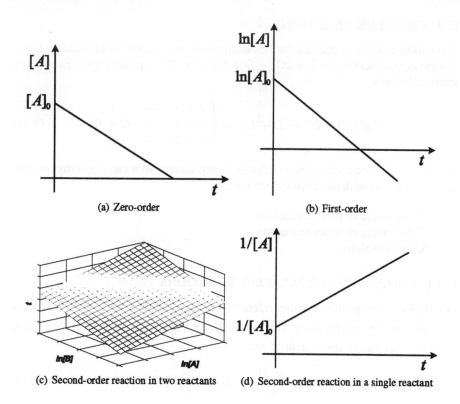

(a) Zero-order

(b) First-order

(c) Second-order reaction in two reactants

(d) Second-order reaction in a single reactant

FIGURE C.1   Evolution over time of mass amounts

and the corresponding solution becomes:

$$\frac{1}{[A]_t} = \frac{1}{[A]_0} + kt \tag{3.12}$$

This solution is shown in Figure C.1(d) (where the line has a positive slope of k). The half-life of this equation is given by:

$$t_{\frac{1}{2}} = \frac{1}{k[A]_0} \tag{3.13}$$

On the other hand, if the reaction is of the form $A + B \rightarrow$ products, the reaction rate will then be defined as:

$$-\frac{d[A]}{dt} = k[A][B] \tag{3.14}$$

and the solution (Fig. C.1(c)) of this differential equation is:

$$ln\frac{[A]}{[B]} = ln\frac{[A]_0}{[B]_0} + ([A]_0 - [B]_0)kt \tag{3.15}$$

C.3   COMPLEX REACTIONS

A complex reaction is one that can be decomposed into a series of elementary steps. For example, reaction $A + B + 2C \rightarrow D + E$ can be decomposed into its elementary steps as follows:

$$(A + B + 2C \rightarrow D + E) \Rightarrow \begin{cases} A + B & \rightarrow & F \\ F + C & \rightarrow & G + D \\ G + C & \rightarrow & E \end{cases} \qquad (3.16)$$

where F and G are reactive intermediates. Complex reactions can generally be categorized into one of the following three groups:

- Competitive or parallel reactions
- Consecutive or series reactions
- Chain reactions

C.3.1   COMPETITIVE OR PARALLEL REACTIONS

A parallel or competitive reaction refers to when a substance reacts simultaneously yielding two distinct products. For two first-order reactions, such as $A \xrightarrow{k_1} B$ and $A \xrightarrow{k_2} C$, the reaction rate is defined as:

$$-\frac{d[A]}{dt} = (k_1 + k_2)[A] \qquad (3.17)$$

The integrated rate equations are then:

$$\begin{cases} [A]_t & = & [A]_0 e^{-(k_1+k_2)t} \\ [B]_t & = & \frac{k_1}{k_1+k_2}[A]_0(1 - e^{-(k_1+k_2)t}) \\ [C]_t & = & \frac{k_2}{k_1+k_2}[A]_0(1 - e^{-(k_1+k_2)t}) \end{cases} \qquad (3.18)$$

A plot of the consumption of $[A]$ and generation of $[B]$ and $[C]$ vs. time t is shown in Figure C.2(a).

C.3.2   CONSECUTIVE OR SERIES REACTIONS

A consecutive or series reaction occurs when a product of one reaction is simultaneously used as a reactant in another reaction. For the following first-order series reaction $A \xrightarrow{k_1} B \xrightarrow{k_2} C$, the reaction rate is defined as:

$$\begin{cases} \text{Reactant A} & \frac{d[A]}{dt} & = & -k_1[A] \\ \text{Reactant B} & \frac{d[B]}{dt} & = & k_1[A] - k_2[B] \\ \text{Product C} & \frac{d[C]}{dt} & = & k_2[B] \end{cases} \qquad (3.19)$$

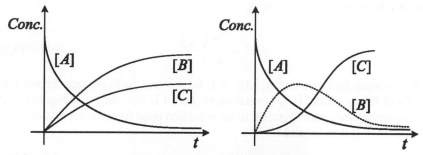

(a) Parallel reaction of two first-order reactants (b) Series reaction of two first-order reactions

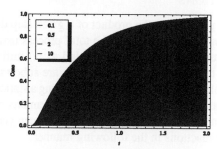

(c) Generation of final product from a series reaction considering different ratios for k_1/k_2

FIGURE C.2 Concentrations *vs.* time

The corresponding integrated rate equations are:

$$
\begin{cases}
[A]_t = [A]_0 e^{-k_1 t} \\[2mm]
[B]_t = \begin{cases} [A]_0 \frac{k_1}{k_2-k_1}(e^{-k_1 t} - e^{-k_2 t}) + [B]_0 e^{-k_2 t} & k_1 \neq k_2 \\[1mm] [A]_0 \, k_1 t e^{-k_1 t} + [B]_0 e^{-k_1 t} & k_1 = k_2 \end{cases} \\[4mm]
[C]_t = \begin{cases} [A]_0 \left(1 + \frac{k_1 e^{-k_2 t} - k_2 e^{-k_1 t}}{k_2 - k_1}\right) + [B]_0(1 - e^{-k_2 t}) + [C]_0 & k_1 \neq k_2 \\[1mm] [A]_0(1 - e^{-k_1 t} - k_1 t e^{-k_1 t}) + [B]_0(1 - e^{-k_1 t}) + [C]_0 & k_1 = k_2 \end{cases}
\end{cases}
$$

$$(3.20)$$

A plot of the consumption of $[A]$, generation/consumption of $[B]$ and generation of $[C]$ vs. time t is displayed in Figure C.2(b).

These equations can be simplified by assuming $k_1 \neq k_2$ and no initial value of $[B]$ and $[C]$ (i.e. $[B_0] = [C_0] = 0$); in this case, the generation of product $[C]_t$ will be:

$$[C]_t = [A]_0 \left(1 + \frac{k_1 e^{-k_2 t} - k_2 e^{-k_1 t}}{k_2 - k_1}\right) \qquad (3.21)$$

with an inflection point occurring at:

$$t^* = \frac{ln\left(\frac{k_1}{k_2}\right)}{k_1 - k_2} \tag{3.22}$$

For example, let's assume that $[A]_0 = 1$, $k_1 = 2$ and the k_1/k_2 ratio to equal 0.1, 0.5, 2 and 10; the plot of $[C]_t$ generation vs. time t is then shown in Figure C.2(c). The respective times corresponding to the inflection point for these various cases are: 0.128 s, 0.347 s, 0.693 s, and 1.279 s.

C.3.3 CHAIN REACTIONS

A chain reaction takes place when the product of a given step during the mechanism is in fact a reactant from a previous step. The total process usually includes various elementary reaction steps, which are often labeled as: a) initiation, b) propagation (or inhibition), and c) termination. Chain reactions are subdivided into two general reaction groups: a) stable, and b) unstable. In considering a stationary (or stable) chain reaction, the concentration of reactive intermediates remains constant over time or slowly decreases. A reaction in the form of $A + B \to F$ can generally be discretized for a set of elementary reactions in the following manner:

$$\begin{cases} \text{Initiation} & A \xrightarrow{k} B + C \\ \text{Propagation}_1 & B + A \xrightarrow{k'} D + F \\ \text{Propagation}_2 & D \xrightarrow{k''} B + E \\ \text{Termination} & 2B \xrightarrow{k'''} G \end{cases} \tag{3.23}$$

The reaction rates for all components are in turn given by:

$$\begin{cases} \frac{d[F]}{dt} & = k'[B][D] \\ \frac{d[B]}{dt} & = k[A] - k'[B][A] + k''[D] - 2k'''[B]^2 \\ \frac{d[D]}{dt} & = k'[B][A] - k''[D] \end{cases} \tag{3.24}$$

In order to solve this (complex) case, it is assumed that the intermediate cases are small enough (i.e. under nearly steady-state conditions):

$$\frac{d[B]}{dt} \approx 0 \tag{3.25}$$

$$\frac{d[D]}{dt} \approx 0 \tag{3.26}$$

Hence, the solution to the differential equations will be:

$$\frac{d[F]}{dt} = k'\left(\frac{k}{2k'''}\right)^{1/2} [A]^{3/2} \tag{3.27}$$

Author Index

ACI-207 247
ACI 318 247
AFNOR P18-454 247
Akhavan, A. 247
Al-Asali, M.M. 259
Alexander, M. G. 248
Alvaredo, A.M. 247
Alves, J.L.D. 250
Amezawa, H. 251
Anon. 247
Arrhenius, S. 247
Asai, Y. 258
ASTM 247
ASTM C1293-08b 247
ASTM C342-97 247
Aziz, T. 260

B., Nooru-Mohamed M. 247
Ballatore, E. 247
Bangert, F. 247
Barin, F.X. 255
Bažant, Z.P. 248
Bažant, Z.P. and Oh, B.H. 248
Bažant, Z.P and Steffens, A. 248
Bawendi, M. 256
Baxter, S. 253
Bay, F. 249
Bazant, Z.P. 248
Becq-Giraudon, E. 247
Ben Haha, M. 248
Bernard, F. 249
Bérubé, M.A. and Duchesnea, J. and
  Doriona, J.F. and Rivestb, M. 248
Bérubé, M.A. and Fournier, B. and
  Bissonnette, B. and Durand, B. 259
Bérubé, M.A. and Smaoui, N. and
  and Fournier, B. and Bissonnette,
  B. and Durand, B. 248
Blaauwendraad, J. 257
Blight, G. 248
Boggs, H.L. 257
Bon, E. 248

Bouchard, P.O. 249
Bourdarot, E. 252, 255, 257, 258
Bournazel, J.P. 248, 258
Broz, J.J. 257
Brühwiler, E. 261

C227, ASTM 247
Capra, B. 248, 257
Carol, I. 248
Carol, I. and Bažant, Z.P. and Prat,
  P.C. 248
Carpinteri, A. 247
CEB 249
Cedolin, L. 249
Cervenka, J. 249
Červenka, V. 249
Chandra, J.M. 249
Charlwood, R. 249
Charlwood, R. G. 249
Charlwood, R.G. 252, 260
Chatterji, S. 249
Chiaramida, A. 249
Chille, F. 248
Clough, R.A. 259
Cognon, H. 257
Comby-Peyrot, I. 249
Comi, C. 249
Constantinera, D. 249
Coppel, F. 249
Cornelissen, H.A.W. 253
Côte, P. 254
Courant, R. 250
Coussy, O. 254, 260
CSA 250
Curtis, D. 250, 251
Curtis, D. D. 249
Curtis, D.D. 260
Cyr, M. 251, 255, 258

D., Kuhl 247
Dahlblom, O. 250
Dauffer, D. 251

Index

297

Printed and bound by CPI Group (UK) Ltd, Croydon, CR0 4YY

18/10/2024

01776267-0006